The Mathematical Principles of Scale Relativity Physics

The Concept of Interpretation

Nicolae Mazilu
Maricel Agop
Ioan Mercheş

CISP

CRC Press
Taylor & Francis Group
Boca Raton London New York

CRC Press is an imprint of the
Taylor & Francis Group, an **informa** business

CRC Press
Taylor & Francis Group
6000 Broken Sound Parkway NW, Suite 300
Boca Raton, FL 33487-2742

First issued in paperback 2021

© 2020 by CISP
CRC Press is an imprint of Taylor & Francis Group, an Informa business

No claim to original U.S. Government works

ISBN 13: 978-1-03-223873-9 (pbk)
ISBN 13: 978-0-367-34934-9 (hbk)

Visit the Taylor & Francis Web site at
http://www.taylorandfrancis.com

and the CRC Press Web site at
http://www.crcpress.com

*"My powers are ordinary.
Only my application brings me success".*

Isaac Newton

Table of Contents

2

Chapter 1. Introduction

Among the newest theories of physics, the Laurent Nottale's scale theory of relativity deserves, in our opinion, a special attention. The scale relativity theory (SRT in what follows) really means business, and big business at that, and we are set here on demonstrating this fact: *SRT targets in fact the very foundations of our positive knowledge.* The proof will be effectively done by showing that SRT follows a line of essential achievements of the physical knowledge of the world, and follows it properly. As a matter of fact the bottom line of our conclusion here is that, once the principle of scale invariance is adopted, there is no other way to follow but the right way, which is the line of thought marked by those essential achievements of knowledge. All of the works to date of Laurent Nottale, regarding the problems raised by scale relativity testify of a well guided thinking, and such a guidance cannot come but from an inherent fundamental principle of knowledge. If there is an ambition from our part here, that would therefore be none other than to make this principle as obvious as possible, maybe even by giving it an explicit verbalization. In doing this, we make use both of common and own results upon the fractal theory of space, expounded though along a special line indicated by Laurent Nottale himself, in an evaluation of thirty years of development of the theory. We quote the final words from a relatively recent book of Nottale:

> Giving up the differentiability hypothesis, i.e. *generalizing the geometric description to general continuous manifolds,* differentiable or not, involves an extremely large number of new possible structures to be investigated and described. In view of the immensity of the task, we have chosen to proceed by steps, *using presently-known physics as a guide.* Such an approach is rendered possible by the result according to which the small scale structures, *which manifest the nondifferentiability,* are smoothed out *beyond some relative transitions toward the large scales.* One therefore recovers

4

the standard classical differentiable theory as a large scale approximation of this generalized approach. But one *also obtains a new geometric theory*, which allows one to understand quantum mechanics as a manifestation of an underlying nondifferentiable and fractal geometry and finally to suggest generalizations of it and new domains of application for these generalizations.

Now the difficulty that also makes their interest with theories of relativity is that *they are meta-theories rather than theories* of *some particular systems*. Hence, after the construction of special relativity of motion at the beginning of the 20 th century, the whole of physics needed to be rendered relativistic (from the viewpoint of motion), *a task that is not yet fully achieved.*

The same is true regarding the program of constructing a *fully scale- relativistic science*. Whatever the already-obtained successes, the task remains huge, in particular when one realizes that *it is no longer only physics that is concerned,* but also *many other sciences*. Its ability to go beyond the frontiers between sciences may be one of the main interests of the scale relativity theory, opening the hope of *a refoundation on mathematical principles* and on predictive differential equations of a philosophy of nature in which physics would no longer be separated from other sciences. [(Nottale, 2011), p. 712; *our Italics*]

The Italics in this excerpt roughly mark our points of intervention with the present work, 'using presently-known... SRT as a guide'. In broad strokes, we aim here to clarify the idea of "general continuous manifolds" and of the general "transition between scales". We also construct a "new geometric theory", with the task of "understanding the quantum mechanics", with a slight change in emphasis: *the quantum mechanics in its wave mechanical form.*

One of the ideas that occurred to us, regarding the fact that theories of relativity are indeed meta-theories rather than theories of some particular systems, is that these theories should in fact not be axiomatically forced upon such systems. It is sufficient that the description of a particular one truly significant for the whole our knowledge be accomplished properly in order to reveal the theory of relativity in it, in all its fundamental features. This shows that the relativity is indeed a meta-theory, by the manner in which it acts as such a meta-theory. Following the usual concept of meta-theory is perhaps the reason why that task of rendering the physics relativistic mentioned by Nottale "is not yet fully achieved". The physical system we have in mind as significant, is significant for the whole modern knowledge indeed: the classical planetary hydrogen atom. Its analysis led to the modern quantum theory, and to a great extent it contributed in the construction of the modern wave mechanics. Nottale has also the great merit of realizing that "no longer only physics" is concerned here. Thus, from the "many other sciences" involved, we shall positively mark in this work the intervention of statistics, theory of color continuum, differential geometry, insofar as it concerns the theory of random surfaces, with a "refoundation on mathematical principles" leading to a sound view of the category of matter, whose sole physical feature is best rendered today by the modern idea of confinement. Perhaps this last observation should be elaborated even from this Introduction a little further.

The critical point of the modern physical knowledge is the one originating in the classical natural philosophy, and due to the essential property of the matter of being presented to our senses through physical structures. Specifically we are talking here of the idea of density, which represents the mathematical characterization of the manner in which *matter fills the space* at its disposal. There is indeed such a manner, insofar as in the physical structures presented to our senses the *matter does not fill the space* suggested to our intellect by their manifestation. In fact, the very concept of a physical structure involves both the theoretical and experimental physics in describing a *specific arrangement of matter in space*, so that the most appropriate definition of a physical structure would be matter penetrated by space. The matter per se shall therefore be defined negatively: *the place where the space does not have access*. This way, the density in its Newtonian connotation can make mathematical sense as it was first defined i.e. by continuity, even by a differential continuity at that *only in the matter thus defined.*

The continuity of matter was always an issue in physics. However, it became a critical issue within the framework of ideas of general relativity, for which the very concept of density is vital, and because of that, in the development of theory one would have to overcome the problems related to the definition of matter as a physical structure. The essential point here is Einstein's idea of describing the world by its mean density. It could not work properly in the background of his general relativity, simply because it is unsecured from a mathematical point of view. Indeed, the theory of general relativity, as a field theory, works mathematically based on exactly the same classical principles of continuity as the classical mechanics. Therefore, as far as the general relativity is concerned, the Newtonian connotation of density cannot be removed. And from this point of view there is not such a precise notion as the mean density, any way we look at it. Quoting Richard Feynman:

> ... What is the situation as regards estimates of the *average density*? If we count galaxies and assume they are more or less like ours, the total density of this kind of visible matter amounts to some $10^{-31} g/cm^3$. This represents some

kind of lower limit on the matter density, since *the visible matter* must be a fraction of the total. ... The critical value, $\rho = 1 \times 10^{-29} g/cm^3$ is always within the range of any estimate; yet the data has enough slop so that if a theory were to require a density as high as $10^{-27} g/cm^3$, the observations could not rule it out, the theory could not be disproved on the basis that the density predicted is much too high.

At this point I would like to make a remark on the present state of observations relevant to cosmology. When a physicist reads a paper by a typical astronomer, he finds an unfamiliar style in the treatment of uncertainties and errors. Although the papers reporting the calculations and measurements are very often very careful in listing and discussing the sources of error, and even in estimating the degree of confidence with which one can make certain key assumptions, when the time comes for quoting a number, such as Hubble time T, one does not find an estimate of the over-all uncertainty, for example in the form of the familiar DT used by physicists. The authors are apparently unwilling to state precisely the odds that their number is correct, although they have pointed out very carefully the many sources of error, and although it is quite clear that the error is a considerable fraction of the number. The *evil* is that often other cosmologists and astrophysicists take this number without regard to the possible error, *treating it as an astronomical observation as accurate as the period of a planet* [(Feynman, 1995), *Chapter 13, our Italics*].

Notice first, that the actual concept of mean density used in general relativity involves the continuity as just a moment of its definition, i.e. through those final figures quoted by Feynman in this excerpt. In its capacity as cosmology, which, as we see it, is actually its essential trait, the general relativity introduces nevertheless another moment of the definition, related explicitly to the manner in which the matter is presented to our senses (*"the visible matter"* in the excerpt above). That moment is the *moment of counting* of the fundamental matter formations, i.e. of the simplest material points presented to our senses. Here, these material points are the galaxies, and *in counting they are reduced to classical material points*, for which the physical structure is totally suppressed. The reason for this situation is way too complicate to be described in a sentence, but we think it comes down to the fact that the universe *appears as isotropic only at the level of galaxies*. There are then many other levels of reasoning in evaluating the mean density of the matter in universe, where the accuracy is lost but, according to Feynman, not assessed in any way, so that it remains in fact completely lost, and the theory based on such evaluation remains a pure speculation. As Feynman articulates it, the "cosmologists and astrophysicists take this number without regard to the possible error", viz. as an exactly defined number, as far as its measure is concerned.

Well, this situation is quite general, it is a human characteristic, we should say: at another level, whatever scientists say, is taken as accurate by the layman, with absolute certainty, because it is... well, scientific! Anyway, Laurent Nottale is an astrophysicist, so that he may have felt 'close range', so to speak, the situation so aptly exposed by Richard Feynman, and perhaps he has decided to clean it up of that 'slop'. Fact is that to Nottale the density can be both the usual physical density and the probability density, and one of the points we shall substantiate here is that this is not a contradiction in terms, provided the matter is properly defined by a *general concept of continuity*. This idea goes far beyond the critique of Feynman who, in chorus with all the physicists and astrophysicists alike, fails to take heed of the fact that the 'average density' itself is only a fictitious concept, improper for the use of general relativity in its classical form, on the very ground that *the matter as presented to our senses does not fill the space of universe*. That failure has an objective reason though, that needs to be uttered from the very beginning: *as long as SRT was not properly developed as a theory, one cannot assess in any way the concept of "average density"*.

Now, before anything else should be said by the way of announcement of intentions and ambitions here, an introductory review of their very origin seems at least appropriate, if not even necessary. This review follows the line of presentation of some specific results of Laurent Nottale himself (Nottale, 1992, 2011). The basis of these results is the SRT's request for a specific fractal structure of the physical theory, which can be recognized first of all in the character of motion at the transition between scales: *space scales* as well as *time scales*. Along this way it is found necessary to describe the physical structure of the matter by a fluid, to wit a complex fluid, complex from quite a few points of view. First, assuming that the curves (fractal, i.e. continuous and non-differentiable) describing the space motions of some fluid particles, are immersed within a *three-dimensional space*, and that the vector $\mathbf{X}$ having components X^i (i = 1,2,3) is the position vector of a point on the fractal curve at the time 't', the total differential expansion up to the third order of a fractal field $F(\mathbf{X}, t, dt)$, with 'dt' the *time resolution scale*, is

$$
\begin{aligned}
d_\pm F = dt \cdot \partial_t F + d_\pm \mathbf{X} \cdot \nabla F + \frac{1}{2} d_\pm X^i d_\pm X^j \cdot \partial_{ij}^2 F \\
+ \frac{1}{6} d_\pm X^i d_\pm X^j d_\pm X^k \cdot \partial_{ijk}^3 F.
\end{aligned}
\tag{1.1}
$$

The sign '+' corresponds to the forward process, while the sign '-' corresponds to the backward one, and we used the abbreviations

$$\partial_{ij}^2 F = \frac{\partial^2 F(\mathbf{X})}{\partial X^i \partial X^j}; \quad \partial_{ijk}^3 F \equiv \frac{\partial^3 F(\mathbf{X})}{\partial X^i \partial X^j \partial X^k}.$$

In equation (1.1) only the terms indicated are finite; any other combinations containing differentials, like $dt^2, dX^i dt, dt^3, dt dX^i dX^j, dt^2 dX^i$ are null when taking an asymptotic limit $dt \to \infty$. We notice that only the first three terms have been used, both in SRT as well as in its nonstandard version (i.e. SRT approach with arbitrary constant fractal dimension).

Take now the forward and backward average values of (1.1). In averaging the fluctuations it is natural to assume that the value of function F and its derivatives are constants at the events location and, moreover, we further assume that the differentials $d_\pm X^i$ and dt are independent. Thus, the average of their products coincides with the product of averages, so that (1.1) becomes

$$d_\pm F = dt \cdot \partial_t F + \langle d_\pm \mathbf{X} \rangle \cdot \nabla F + \frac{1}{2} \langle d_\pm X^i d_\pm X^j \rangle \cdot \partial_{ij}^2 F$$
$$+ \frac{1}{6} \langle d_\pm X^i d_\pm X^j d_\pm X^k \rangle \cdot \partial_{ijk}^3 F. \tag{1.2}$$

Using by now the standard relations

$$d_\pm \mathbf{X} \equiv d_\pm \mathbf{x} + d_\pm \boldsymbol{\xi}, \tag{1.3}$$

where $d_\pm \mathbf{x}$ is differentiable and resolution scale independent spatial coordinate variation, and $d_\pm \boldsymbol{\xi}$ is non-differentiable (fractal) and resolution scale dependent spatial coordinate variation, we get:

$$d_\pm F = dt \cdot \partial_t F + \langle d_\pm \mathbf{x} \rangle \cdot \nabla F$$
$$+ \frac{1}{2} (d_\pm x^i d_\pm x^j + \langle d_+ \xi^i d_+ \xi^j \rangle) \cdot \partial_{ij}^2 F \tag{1.4}$$
$$+ \frac{1}{6} (d_\pm x^i d_\pm x^j d_\pm x^k + \langle d_\pm \xi^i d_\pm \xi^j d_\pm \xi^k \rangle) \cdot \partial_{ijk}^3 F.$$

One can recognize here that even if the average value of the fractal coordinate differential is null, the situation can still be different for a higher order of fractal coordinate average. Indeed, let us first focus on the averages of second order terms. If $i \neq j$, these averages are zero due to the independence of the increments involved in them. Thus, using the fractal equations we have:

$$\langle d_\pm \xi^i \rangle = \lambda_\pm^i \left(\frac{dt}{\tau} \right)^{1/D_f}, \tag{1.5}$$

9

where $\lambda_{\pm}^i$ are constant coefficients having statistical meanings, dt/τ is the normalized resolution scale, with τ the time scale and D_f is fractal dimension of the motion curve. Thus, we can write:

$$\langle d_{\pm}\xi^i d_{\pm}\xi^j \rangle = \lambda_{\pm}^i \lambda_{\pm}^j \left(\frac{dt}{\tau} \right)^{2/D_f}. \tag{1.6}$$

Then, let us consider the averages of third degree. If $i \neq j \neq k \neq i$, these averages are zero due to the independence of the variations of the first order. Now, using (1.5) we get:

$$\langle d_{\pm}\xi^i d_{\pm}\xi^j d_{\pm}\xi^k \rangle = \lambda_{\pm}^i \lambda_{\pm}^j \lambda_{\pm}^k \left(\frac{dt}{\tau} \right)^{3/D_f}. \tag{1.7}$$

For the fractal dimension D_f any definition can be used (the Hausdorff-Besikovich fractal dimension, the Kolmogorov fractal dimension etc.), but once a definition accepted, it has to be maintained the same over the entire analysis of the complex fluid dynamics. Thus, (1.4) may be simplified by inserting (1.5), (1.6) and (1.7), dividing by 'dt' and neglecting the straight differentials of higher orders. The net result is:

$$\frac{d_{\pm}F}{dt} = \partial_t F + \mathbf{V}_{\pm} \cdot \nabla F + \frac{1}{2}\frac{\lambda_{\pm}^i \lambda_{\pm}^j}{\tau} \cdot (\partial_{ij}^2 F) \cdot \left(\frac{dt}{\tau} \right)^{(2/D_f)-1}$$
$$+ \frac{1}{6}\frac{\lambda_{\pm}^i \lambda_{\pm}^j \lambda_{\pm}^k}{\tau} \cdot (\partial_{ijk}^3 F) \cdot \left(\frac{dt}{\tau} \right)^{(3/D_f)-1}. \tag{1.8}$$

Now, recalling that the fractal operator of time differentiation has the form:

$$\frac{d}{dt} = \frac{1}{2}\frac{d_+ + d_-}{dt} - \frac{i}{2}\frac{d_+ - d_-}{dt}, \tag{1.9}$$

the velocity of the process becomes a complex vector

$$\mathbf{V} \equiv \frac{d\mathbf{X}}{dt} = \frac{1}{2}\frac{\mathbf{V}_+ + \mathbf{V}_-}{dt} - \frac{i}{2}\frac{\mathbf{V}_+ - \mathbf{V}_-}{dt} = \mathbf{V}_d - i\mathbf{V}_f. \tag{1.10}$$

The real part $\mathbf{V}_d$ of the complex velocity $\mathbf{V}$ represents the *standard classical velocity*, which does not depend on the resolution scale, while the imaginary part $\mathbf{V}_f$ is a new quantity coming from *non-differentiability* (fractal velocity), and is *dependent on the resolution scale*. Moreover, applying the fractal operator (1.9) to the fractal field $F(\mathbf{X}, t, dt)$ by means of relations (1.8) we have:

$$\frac{dF}{dt} = \partial_t F + \mathbf{V} \cdot \nabla F + \frac{1}{4\tau}[(\lambda_+^i \lambda_+^j + \lambda_-^i \lambda_-^j)$$

$$-i(\lambda_+^i \lambda_+^j - \lambda_-^i \lambda_-^j)] \cdot (\partial_{ij}^2 F) \cdot \left(\frac{dt}{\tau}\right)^{(2/D_f)-1}$$

$$+\frac{1}{12\tau}[(\lambda_+^i \lambda_+^j \lambda_+^k + \lambda_-^i \lambda_-^j \lambda_+^k)$$

$$-i(\lambda_+^i \lambda_+^j \lambda_+^k - \lambda_-^i \lambda_-^j \lambda_+^k)] \cdot (\partial_{ijk}^3 F) \cdot \left(\frac{dt}{\tau}\right)^{(3/D_f)-1},$$

(1.11)

which then allows us to give an explicit form of the fractal operator (1.10):

$$\frac{d}{dt} = \partial_t + \mathbf{V} \cdot \nabla$$

$$+\frac{1}{4\tau}\left(\frac{dt}{\tau}\right)^{(2/D_f)-1}[(\lambda_+^i \lambda_+^j + \lambda_-^i \lambda_-^j) - i(\lambda_+^i \lambda_+^j - \lambda_-^i \lambda_-^j)] \cdot \partial_{ij}^2$$

$$+\frac{1}{12\tau}\left(\frac{dt}{\tau}\right)^{(3/D_f)-1}[(\lambda_+^i \lambda_+^j \lambda_+^k + \lambda_-^i \lambda_-^j \lambda_+^k)$$

$$-i(\lambda_+^i \lambda_+^j \lambda_+^k - \lambda_-^i \lambda_-^j \lambda_+^k)] \cdot \partial_{ijk}^3.$$

(1.12)

For *Markov-type random processes*, in which case we have

$$\lambda_+^i \lambda_+^j = -\lambda_-^i \lambda_-^j = 2\lambda^2 \delta^{ij};$$

$$\lambda_+^i \lambda_+^j \lambda_+^k = -\lambda_-^i \lambda_-^j \lambda_-^k = 2\sqrt{2}\lambda^3 \delta^{ijk},$$

(1.13)

the fractal operator (1.12) takes the form:

$$\frac{d}{dt} = \partial_t + \mathbf{V} \cdot \nabla - i\frac{\lambda^2}{\tau}\left(\frac{dt}{\tau}\right)^{(2/D_f)-1}\nabla^2$$

$$+\frac{\sqrt{2}}{3}\frac{\lambda^3}{\tau}\left(\frac{dt}{\tau}\right)^{(3/D_f)-1}\nabla^3$$

(1.14)

where ∇^2 is the usual Laplace operator, and we adopted the notations:

$$\nabla^3 = \frac{\partial^3}{\partial x^3} + \frac{\partial^3}{\partial y^3} + \frac{\partial^3}{\partial z^3};$$

$$\delta^{ij} \equiv \begin{cases} 1 & i = j \\ 0 & i \neq j \end{cases}; \qquad \delta^{ijk} \equiv \begin{cases} 1 & i = j = k \\ 0 & i \neq j \neq k. \end{cases}$$

According to the regular theory the first three terms of the the fractal operator (1.14) correspond to *a covariant derivative in the fractal space*. Extending this interpretation to all of the terms of operator represented in equation (1.14),

11

warrants a corresponding extension of the results, which we have generally described as follows [see (Agop, Păun & Harabagiu, 2008); (Casian-Botez, Agop, Nica, Păun & Munceleanu, 2010); (Mercheş & Agop, 2015)].

First, the principle of scale covariance was applied and, accordingly, we have assumed that the transition from classical (differentiable) to "fractal" physics can be implemented by replacing the standard time derivative operator (d/dt), with the fractal operator (d/dt). Consequently, by applying the fractal operator (1.14) to the complex velocity field **V**, the specific momentum conservation law in its covariant form (i.e. the equation of geodesics) can be written as

$$\frac{d\mathbf{V}}{dt} = \partial_t \mathbf{V} + (\mathbf{V} \cdot \nabla)\mathbf{V} - i\frac{\lambda^2}{\tau}\left(\frac{dt}{\tau}\right)^{(2/D_f)-1}\nabla^2\mathbf{V}$$
$$+ \frac{\sqrt{2}}{3}\frac{\lambda^3}{\tau}\left(\frac{dt}{\tau}\right)^{(3/D_f)-1}\nabla^3\mathbf{V} = \mathbf{0}. \tag{1.15}$$

Equation (1.15) shows that at any point of a fractal path, the *local acceleration* $\partial_t \mathbf{V}$, the convection $(\mathbf{V} \cdot \nabla)\mathbf{V}$, the *dissipation* $(\lambda^2/\tau)(dt/\tau)^{(2/D_f)-1}(\nabla^2\mathbf{V})$, and the *dispersion* $(\lambda^3/\tau)(dt/\tau)^{(3/D_f)-1}(\nabla^3\mathbf{V})$, are in a sort of equilibrium. The presence of the dissipative and dispersive terms in equation (1.15) shows that the behavior of the complex fluid is of viscoelastic, or even hysteretic type. Since there are no interactions in the fluid, when it is assimilated with a fractal fluid, the *self-convection, self-dissipation*, and *self-dispersion* type mechanisms should be operational. Therefore, in this case, the geodesics of the fractal space can be assumed to correspond to the streamlines of a complex fluid of free particles, thereby allowing us to introduce a specific mathematical treatment of the problem. *The motion thus described is purely inertial.*

Now, let us consider only the *dissipative* approximation of motions on fractal paths (the dispersive term from equation (1.15) is taken as negligible comparing to the dissipation and convection terms). Then the equation (1.15) is:

$$\frac{d\mathbf{V}}{dt} = \partial_t \mathbf{V} + (\mathbf{V} \cdot \nabla)\mathbf{V} - i\frac{\lambda^2}{\tau}\left(\frac{dt}{\tau}\right)^{(2/D_f)-1}\nabla^2\mathbf{V} = \mathbf{0}. \tag{1.16}$$

For reasons to be explained shortly, one chooses **V** in the form:

$$\mathbf{V} = -2i\frac{\lambda^2}{\tau}\left(\frac{dt}{\tau}\right)^{(2/D_f)-1}\nabla\ln\Psi. \tag{1.17}$$

Substituting (1.17) in (1.16), one can get after calculations

$$\frac{\partial\Psi}{\partial t} - i\frac{\lambda^2}{\tau}\left(\frac{dt}{\tau}\right)^{(2/D_f)-1}\nabla^2\Psi = (\nabla \times \Phi)\Psi, \tag{1/18}$$

12

where Φ is an arbitrary function of position. Relation (1.18) is the reason of the previous choice for the field of velocities $\mathbf{V}$: *we have to do with a nonstationary Schrödinger-type equation.*

The potential $(\ln\Psi)$ was originally introduced for purposes of description of the dissipation in wave mechanics (Kostin, 1972). Remarkably enough, it describes the very properties of the free particle within fractal mechanics as conceived here, and these turn out to spring actually from a wave mechanics. As a matter of fact, the quantization of Schrödinger equation with pure dissipation terms is known to lead to Gaussian wave packets (Hasse, 1978). But, again, we follow another path, having an objective connotation related to hydrodynamics. Namely, we can solve the Schrödinger equation by the classical methods that reduce it to the equation of continuity and Navier-Stokes equation (the so-called Madelung representation), thus giving an explicit form to the self-interaction. So we can handle the problem by known techniques (Bohm, 1952).

Specifically, for $\Psi = \sqrt{\rho} \cdot e^{iS}$, with $\sqrt{\rho}$ the amplitude and S the phase of ψ, the complex velocity field (1.17) takes the explicit form:

$$\mathbf{V} = \frac{\lambda^2}{\tau} \left(\frac{dt}{\tau} \right)^{(2/D_f)-1} (2\nabla S - i\nabla\ln\rho) \equiv \mathbf{V}_d - i\mathbf{V}_f \qquad (1.19)$$

with an obvious identification of the two real components of the complex velocity field. Now carrying the calculations through, we find the interesting result that the quantum potential is entirely determined by the fractal component of the velocity field:

$$\frac{\partial \mathbf{V}_d}{\partial t} + (\mathbf{V}_d \cdot \nabla)\mathbf{V}_d = -\nabla Q; \qquad \frac{\partial \rho}{\partial t} + \nabla \cdot (\rho \mathbf{V}_d) = 0 \qquad (1.20)$$

with Q the specific fractal potential:

$$\begin{aligned} Q &= -2\frac{\lambda^4}{\tau^2} \left(\frac{dt}{\tau} \right)^{(4/D_f)-2} \frac{\nabla^2 \sqrt{\rho}}{\sqrt{\rho}} \\ &= -\frac{\mathbf{V}_f^2}{2} - \frac{\lambda^2}{\tau} \left(\frac{dt}{\tau} \right)^{(2/D_f)-1} (\nabla \cdot \mathbf{V}_f). \end{aligned} \qquad (1.21)$$

The first equation (1.20) represents the conservation law for momentum, while the second equation represents the conservation law of a probability density. Let us notice that when multiplying the probability density with the rest mass of the complex fluid particle, we obtain its standard density (Nottale, 1992, 2011). Then, the second equation (1.20) is a continuity equation that corresponds to the complex fluid continuity law. Through the fractal velocity field $\mathbf{V}_f$, the specific fractal potential Q is a measure of non-differentiability of the complex fluid particle trajectories, more to the point, a measure of their chaoticity. The equations (1.20) with (1.21) define a non-differentiable hydrodynamic model.

Since the position vector of the particle is assimilated with a stochastic process of Wiener type, ψ is not only the scalar potential of a complex velocity (through $\ln\psi$) in the framework of fractal hydrodynamics, but also the basis of a density of probability (through $|\psi|^2$) in the framework of a Schrödinger type theory. Thus we have a complementarity of these two formalisms (the formalism of the fractal hydrodynamics and the one of the Schrödinger type equation). Moreover, the chaoticity, either through turbulence in the fractal hydrodynamics approach, or through stochasticization in the Schrödinger type approach, is generated only by the non-differentiability of the movement trajectories in a fractal space. Thus, it can be shown that the dissipation-dominated fractal behaviour of a fluid-solid mixture, for instance, can be properly described by the Schrödinger equation for the free particle (Nedeff, Lazăr, Agop, Eva, Ochiuz, Dimitriu, Vrăjitoriu & Popa, 2015). This is another way of saying that the part of this behavior of fractal fluid is purely inertial.

Now, let us concentrate upon the cases where the dissipation is negligible, however the third order term in the fractal expansion (1.19) is somehow dominant. The fractal equation (1.15) becomes:

$$\frac{d\mathbf{V}}{dt} = \partial_t \mathbf{V} + (\mathbf{V} \cdot \nabla)\mathbf{V} + \frac{\sqrt{2}}{3} \frac{\lambda^3}{\tau} \left(\frac{dt}{\tau} \right)^{(3/D_f)-1} \nabla^3 \mathbf{V} = \mathbf{0}. \quad (1.22)$$

The behavior of fractal fluid described by this equation seems to us to be of a *dispersion* type. In thus labelling the fractal fluid, we took as a prototype of this kind of behavior the third order Korteweg-de Vries equation, describing dissipation-free waves *dominated by dispersion* in one space dimension. This equation is known to have solutions preserving their space-time shape through interaction (Gardner, Greene, Kruskal, & Miura, 1967). These are particle-like solutions, or solitons (Scott, Chu, & McLaughlin, 1973). In such a circumstance, our fractal expansion leads explicitly to a third order nonlinear equation and obviously also to a soliton solution. It is this fact only that induced us into saying that the equation (1.22) describes a *dispersive behavior* of the fractal fluid. By comparison with the previous results, one can expect that, although in this formalism the particle and soliton solutions involve the same inertial type of behaviour − the motion is still geodesic − the soliton is structurally expected to exhibit a new characteristic. It turns out that this is a kind of holographic property: *any volume of the fluidic soliton has the same structure as the entire fluid.*

This brief review appears as just enough in order to support what we have called before our statement of intentions and ambitions with the present work. The general conclusion sustained by SRT along the previous lines initiated by

Laurent Nottale, is that the description of the world is legitimate as *a fractal fluid of free particles*. However, this fractal fluid should be described via Schrödinger equation in its *both instances* of time dependent as well as time independent partial differential equation. Our proper finding at this juncture was that, while the second order partial differential equation describes a *dissipation* indeed, a third order partial differential equation of Schrödinger type is instrumental in describing a *dispersion* within the fractal fluid.

Now, at some point of our research an occurrence has prompted a certain amount of thinking from our part, about the difference between time-dependent and time-independent Schrödinger equations. Namely the solitonic solutions of the third order partial differential equation are also formally to be found among some solutions of the second order, even linear, partial differential equation, or vice versa, does not really matter. What is important here, is the fact that between the two different 'approximations', as we called them, of the fractal Schrödinger equations, there seem to be some fundamental connections. If discovered, these connections might be able to draw SRT out of the pit of conjectures, giving it the place of a fundamental theory, which we think it plainly deserves. Thus, our main intention here, inspired basically by this last finding of ours along the Nottale's line of thought, is to show that SRT *is in fact a unitary view of matter*. From physical point of view, the present work is, by and large, set out to discuss the actuality of this statement.

The first thing to prove along this line is that the wave-mechanical description of the matter is universal, not just incidental. Specifically, the time-dependent Schrödinger equation is universal: it is the equation representing what is currently known in the theory of elementary particles as the principle of asymptotic freedom, but in any structure of the matter, at any scale [see (Nottale, 2011), Open Problem 22]. The only thing incidental at this theoretical level is a Born-type interpretation of the wave function. The wave function per se is a universal instrument of knowledge, transcending between any two scales, particularly between quotidian world and cosmological scales of the theoretical physics (Mazilu & Porumbreanu, 2018). The potential, on the other hand, is well known to be related to a wave function through the time-independent or stationary Schrödinger equation (Nottale, 1992). However, the wave function of this last instance is a very special one: it is the one obtained *from the density of matter, by square rooting*. Therefore this is the function for which the Born's probabilistic interpretation may be valid, and this property goes for no other wave function. The proper conclusion would then be that SRT is, indeed, a theory of matter structures in the universe, *provided one goes beyond the Born-like probabilistic interpretation of the wave function*. A general probabilistic approach then emerges quite naturally, by giving Nottale's ideas the rank of principles of knowledge which they plainly deserve.

Perhaps a few details of the organization of the present work will be useful, and we really think they are indeed, at least for a chance reader. The first order of things is triggered by the fluid dynamical approach of the wave mechanics: the structure thus revealed through the Schrödinger equation is that of a *Madelung fluid*, that occurred with a certain, allegedly physical, interpretation of the wave function (Madelung, 1927). This is undoubtedly at odds with the Born interpretation of the wave function, but as we shall see, it allows us to restore what has been lost by the acceptance of Born-type stand. The moment can be aptly characterized by the words of Henri Bacry:

> 2. *All particles are equal*
>
> One of the important ideas of the century was that of de Broglie, who proposed in 1923 *to put the matter and radiation on the same footing*. An irony of the history was that this democracy *was destroyed by the Born statistical interpretation of the wave function*, an interpretation *that the photon field cannot have*. Obviously, such an interpretation *was a non-relativistic property*, but it was believed that it was valid also for relativistic waves.
>
> With our proposal, *the symmetry between all kinds of particles is recovered*. First, each kind of particle has a position operator. *All spinning particles have in common the property of not being localizable*. Since all stable particles are spinning, this *provides the spin with a fundamental character*.
> [(Bacry, 1988); *our Italics*]

We will show in a section of the present work that the properties which Louis de Broglie revealed in that 'photon field' alluded to by Bacry in this excerpt, *should be assumed at any rate for a properly defined wave function*, and this request cannot be satisfied but only by *free particles*. Then again, we seem to come at odds with that 'proposal' of Henri Bacry regarding the position operator, inasmuch as it involves interactions, and this certainly means no free particles. One section here is therefore dedicated to this issue, where it is shown that the fluid mechanics in the Madelung's taking, allows a meaningful interpretation of the position variables. This interpretation is of a purely classical origin, thoroughly pointed out and analyzed in the due time and space of the present work.

Thus, we come up with a physical model of the world matter *per se*, the matter free of space: a fluid with no interaction, only inertially described, just the fluid necessary for SRT. For once, *this is not a physical structure*, but only a physical model, because the matter does not allow for penetration of the space within it. And only if we have matter penetrated by space can we say that we are dealing with a physical structure, otherwise not. Therefore, one of the sections of the present work will be dedicated to the explanation of the inertia and its mathematical extension. The explanation will be done on purely geometrical grounds, whereby the holographic principle - a natural principle, we should say, for the matter in bulk - is brought to bear at the level of a surface, where the surface is this time physical: *the surface of separation of matter from space.* The physical definition of such a surface is by fragments, representing portions of matter just about to enter the reality of a physical structure. This plainly gives physical sense to holographic principle in its most general undertaking.

Finally, we feel necessary to round up with the physical grounds of the SRT. Suffice it to mention for now two issues of principle. First we show why de Broglie's approach to democracy is universal, by the way of discussing a fundamental model of the natural philosophy: *the planetary model* of atom. From this we extract the idea of elementary particle of modern physics. Secondly, we describe the basis of a general mathematical approach of scale relativity, thus of *resolution scale*: this is just the scale making the finite... 'finite' in any physical world, be it microcosmos, quotidian cosmos or the universe at large.

Chapter 2. Madelung Fluid Dynamics

Erwin Schrödinger left a marvelous heritage for the future of physics, namely a way to look at the spatially extended systems of known physical components: not each component in evolution, as in classical physics, but all of the components at once; simultaneously, as it were, when we can actually speak of simultaneity here. The mathematical device through which this approach is accomplished is the well-known *Schrödinger equation*, to be satisfied by what came ever since to be known as the *wave function*. Needless to say, the initial success of the method was purely mathematical — the presentation of the planetary atom as an eigenvalue problem — and thus left many open issues, especially on the physical side of the problem at hand. Among these issues, the *physical interpretation* of the wave function is a chief one, which still haunts the natural philosophy today. Almost immediately after the fundamental works of Schrödinger, Max Born and Erwin Madelung came up with two physical interpretations of the wave function, one from the point of view of the theory of probabilities, the other from the point of view of the classical fluids' dynamics. Momentarily we are concerned with Madelung's physical interpretation, as Born interpretation of the wave function seems to be only incidental in the ultimate analysis we intend to pursue in this work.

Erwin Madelung has noticed that the task of interpretation is twofold: first of all we have to give a *physical* interpretation to the stationary equation

$$\frac{\hbar^2}{2m}\Delta\Psi_0 + (W - U)\Psi_0 = 0; \qquad \Psi = \Psi_0 e^{i\frac{W}{\hbar}t}$$

obtained by Schrödinger via a logarithmic transformation of the classical variable of action, which helped transform the energy of the planetary atom into a variational functional of Dirichlet type (Schrödinger, 1933). The equation above is then the corresponding partial differential equation obtained from that variational principle. Then Madelung realized that in order to give a proper physical interpretation to the function Ψ itself which brings that logarithmic transformation of Schrödinger into effect one needs actually to give a physical interpretation to the nonstationary equation that Schrödinger obtained from stationary equation by eliminating time with the help of the function Ψ itself. That elimination of time is generally a matter of arbitrariness, forasmuch as the derivative of any order of a wave function which depends on time through a phase factor with the phase linear in time, is proportional to the very function. Schrödinger has probably had a long debate with himself over the choice of the right way of doing it. First, he obviously inclined for a wave equation of the D'Alembert (Klein-Gordon) type, but soon he realized that the analogy with classical problem of the membrane fits better his own philosophy, and allows for an equation equivalent with the first order derivative elimination of time. The equation obtained this way came to be known as the nonstationary (or time- dependent) Schrödinger equation. From this perspective, the way of Madelung's physical interpretation can pe summarized along the following modern lines.

The Madelung Fluid

Let us consider a *continuous material system*, described by a density ρ, function of position and time in a certain frame of reference, and normalized such that

$$\iiint_\Omega \rho(\mathbf{x}, t) d^3\mathbf{x} = 1. \tag{2.1}$$

Here Ω is the spatial range of extension of that continuous matter. Assume now that the system thus chosen can also be described by a wave function Ψ, solution of the time-dependent Schrödinger equation:

$$i\hbar \frac{\partial \Psi}{\partial t} = -\frac{\hbar^2}{2m} \Delta \Psi + V(\mathbf{x}, t)\Psi. \tag{2.2}$$

Here $V(\mathbf{x}, t)$ is the potential of forces, as it appears in classical physics, assumed a function of position and time. The problem arises if the two descriptions are compatible and in what conditions. In view of (2.1), the density of this continuous system *can also be taken as a probability density*, so that a connection can be *a priori* posited between the two descriptions. At this moment Madelung assumes what almost any modern approach assumes, namely a Born-type correlation between wave function and the density of fluid:

$$\rho(\mathbf{x}, t) = |\Psi(\mathbf{x}, t)|^2. \tag{2.3}$$

The problem can now be transferred into that of analyzing what this correlation entails. Note that the equation (2.3) entitles one to define the wave function as an *essentially complex* function:

$$\Psi(\mathbf{x}, t) = R(\mathbf{x}, t)e^{iS(\mathbf{x}, t)}, \qquad R(\mathbf{x}, t) = \sqrt{\rho(\mathbf{x}, t)} \tag{2.4}$$

19

where S is, for the moment, an arbitrary function of position and time. Then the Schrödinger equation (2.2) may be translated into a system of two partial differential equations:

$$\frac{\partial R}{\partial t} = -\frac{\hbar}{m}[R\nabla^2 S + 2\nabla R \cdot \nabla S],$$

$$-\hbar R\frac{\partial S}{\partial t} = VR - \frac{\hbar^2}{2m}[\nabla^2 R - R(\nabla S)^2].$$

(2.5)

The first of these equations can be cast into:

$$\frac{\partial \rho}{\partial t} + \frac{\hbar}{m}\nabla \cdot (\rho\nabla S) = 0$$

(2.6)

provided R is not identically zero. By this last equation the matter thus described gets actually a *physical structure*: the equation represents the continuity equation for a *continuous fluid* of density ρ, function of position and time. This physical structure is represented by *an ensemble of classical material points* having a velocity field mathematically defined by the expression:

$$\mathbf{v}(\mathbf{x}, t) = \frac{\hbar}{m}\nabla S$$

(2.7)

which shows that S should be the *classical action* function. This ensemble is, in fact, the key of *interpretation* which Erwin Madelung gives to the Schrödinger's wave function. The further connection of this velocity field with the wave function itself can be given by using the definition (2.4), with the important result:

$$\mathbf{v}(\mathbf{x}, t) = -\frac{i\hbar}{2m}\nabla\ln\left(\frac{\Psi}{\Psi^*}\right)$$

(2.8)

the star denoting the complex conjugate of the function. Thus the equation (2.6) is nothing more than the condition of conservation of mass on a continuous ensemble of classical material points, characterized by the density ρ, function of position and time.

On the other hand, we need to consider the second equation of the system (2.5), which can be transcribed into the form:

$$\hbar\frac{\partial S}{\partial t} + \frac{\hbar^2}{2m}(\nabla S)^2 + V - \frac{\hbar^2}{2m}\frac{\nabla^2\sqrt{\rho}}{\sqrt{\rho}} = 0.$$

(2.9)

As S has already been identified with the classical action, this equation is a Hamilton-Jacobi equation, having however the potential changed by a quantity

20

depending on the *space fluctuations of the density,* and represented specifically in the form:

$$Q(\mathbf{x}, t) \equiv -\frac{\hbar^2}{2m} \frac{\nabla^2 \sqrt{\rho}}{\sqrt{\rho}} \qquad (2.10)$$

which contains the square of the Planck's constant. This is the well-known 'quantum contribution' to potential, discovered by Erwin Madelung as a consequence of his interpretation. On the other hand, the existence of this contribution is a condition that the physics of the Madelung fluid should be consistent. It is here the point where Madelung's mathematical treatment basically stops, giving the floor to the discussion of some questions raised by the alternatives of the physical interpretation of the theory. In his own words, these alternatives are summarized in the following excerpt closing the Madelung's original work:

> We probably have to decide only between the following alternatives:
>
> a) Do several electrons flock together to form a larger entity?
>
> b) Do they exclude each other, and interfere (only) under certain conditions?
>
> c) Do they penetrate each other without merging?
>
> It seems to me that c) is the most likely. a) would mean the same solution like that for a single electron, *only with an altered normalization,* which obviously would lead to false results. b) seems unlikely from the perspective of "submerged orbits", *but still conceivable.*
>
> According to c) *several vectors must be defined in each point of space,* as well as the corresponding velocity potentials. The continuum would then have an appearance suggesting a vivid swarm, *whose particles have infinite free paths.*
>
> *What form the function U assumes, representing both the interaction between electrons and the "quantum contribution" from equation* (2.10), can be decided only after the successful calculations for at least one case.
>
> There is thus a *prospect of building a quantum theory of the atom* on this basis. The *radiation emission* is thereby only partially represented. To be sure, it appears that the atom in a quantum state does not radiate, and also the frequencies of the radiation are shown correcly, only without "jump", but rather *at a slow transition into a state of non-stationarity;* but many other things such as, for instance, *the absorption of quanta,* remain unclear. [(Madelung, 1927); *our translation and Italics*]

The story of accomplishments related to the lines of question raised by Erwin Madelung in this excerpt, can be briefly reviewed as in the immediately following lines. We do not follow the historical order, by any means, not even the order of the questions as raised by Madelung himself, because we are interested only in physical achievements. As it turned out, almost every single one of those issues - the Italicized ones for sure! - have been brought about in a way or another in theoretical physics today

First, denoting $U \equiv V + Q$, as that potential «representing both the interaction between electrons and the "quantum contribution"», applying the gradient operation to (2.9), and then multiplying the result with the density ρ, gives the equation

$$\frac{\partial}{\partial t}(m\rho v_j) + \frac{\partial}{\partial x^k}(m\rho v_k v_j) + \rho\frac{\partial U}{\partial x^j} = 0 \qquad (2.11)$$

which is a *transport equation* for the momentum of the Madelung fluid, the so-called *Navier- Stokes equation.* Starting from this, H. E. Wilhelm performed some of those 'successful calculations' for the case of hydrogen atom, giving then a hydrodynamical interpretation of the results (Wilhelm, 1971). Wilhelm follows exactly the line of Madelungs's ideas, forasmuch as he uses the Schrödinger equation in both its instances (stationary and nonstationary) on the same footing, in order to construct the wave function, from which then he finds the density of the corresponding Madelung fluid.

But the things turned out to be a lot more complicated than the classical mechanics would have been able to indicate. Specifically, the potential Q is different from the classical potential V not only by the fact that it contains the Planck's constant, but also by the fact that the *quantum force* resulting from it by the classical operation of gradient, is in fact the *divergence of a tensor*. This is not to say that a gradient of a scalar function cannot be the divergence of a tensor, but in keeping with the classical views here, that tensor should be a *stress tensor*, and this is what brings all the complications. Indeed, customarily — which means *within the classical theory of the forces* in a continuum — the gradient of a potential like Q from equation (2.10) should represent a force. As the gradient of V is a known conservative force, at least in the case of the classical planetary model, with the classical part of the potential we have no issues here. However, the issues occur once we consider the 'quantum contribution' to potential. Specifically, if we take the tensor [see (Takabayasi, 1952), p. 180; (Wilhelm, 1971)]:

$$\tau_{ij} = \frac{\hbar^2}{2m}\rho\frac{\partial^2}{\partial x^i \partial x^j}(\ln\rho) \qquad (2.12)$$

the equation defining the forces that derive from potential Q can be written in the form of an equilibrium equation, physically characterizing our continuous medium – the Madelung fluid – by tensions incorporated in the tensor , and acted upon by volume forces included in the gradient of the entire potential (∇Q):

$$\rho \frac{\partial Q}{\partial x^i} + \frac{\partial \tau_{ij}}{\partial x^j} = 0. \tag{2.13}$$

This proves our statement. Part of the complications are then brought about by this very observation, because in *every physical fluid a constitutive law should be in effect*, relating the acting stress to the deformation of the fluid. And the Madelung fluid can be taken, indeed, as a physical fluid, for in fact one can define a velocity field by the variation of density:

$$\mathbf{u}(\mathbf{x}, t) \equiv \frac{\hbar}{m\sqrt{2}} \frac{\nabla \rho}{\rho} \tag{2.14}$$

so that the tensor τ_{ij} can be written in the form:

$$\tau_{ij} = \eta \cdot \left(\frac{\partial u_i}{\partial x^j} + \frac{\partial u_j}{\partial x^i} \right), \qquad \eta \equiv \frac{\hbar}{2\sqrt{2}} \rho. \tag{2.15}$$

This is, indeed, a linear constitutive equation for a viscous fluid, and gives the reason for an original interpretation of the coefficient η as a *dynamic viscosity* of the Madelung fluid (Harvey, 1966). Summarizing the results up to this point, according to Madelung's idea, *the physical system* allowing a proper interpretation of the wave function via transformation (2.4) should be *a viscous fluid having a density proportional with the square of modulus of the wave function.*

The first problem to be now solved is to decide if the velocity field $\mathbf{u}$ as defined by equation (2.14) is indeed of a quantum nature, i.e. if it necessarily contains the Planck's constant. Especially Nathan Rosen reiterated quite a few times this issue, starting from the fifth decade of the last century (Rosen, 1945, 1947). The equation (2.9) shows that the function S is not quite as arbitrary as one may think, merely based on mathematics: it should be the classical action, or at least a function having some relation to the classical action. Thus, writing the exponent of the wave function from (2.4) in a manifestly nondimensional form, as in fact an exponent should mathematically be, we need to put out $(S/\hbar)$ instead of S. So we can consider the function S as having its physical meaning, which makes obvious that the continuity equation (2.6) and the vectorial field $\mathbf{v}$ from (2.7) are of a purely classical extraction, as they do not contain the Planck's constant:

$$\frac{\partial \rho}{\partial t} + \nabla \cdot (\rho \mathbf{v}) = 0, \quad m\mathbf{v}(\mathbf{x}, t) \equiv \nabla S. \tag{2.16}$$

At the same time, though, in equation (2.9) the potential Q remains unchanged, which means that it is of a purely quantal nature:

$$\frac{\partial S}{\partial t} + \frac{1}{2m} (\nabla S)^2 + V = \frac{\hbar^2}{2m} \frac{\nabla^2 \sqrt{\rho}}{\sqrt{\rho}}. \tag{2.17}$$

23

Consequently the velocity field **u** itself, as defined by (2.13), is indeed of a purely quantal nature. However, this equation makes clear that we can have a classical fluid only in the cases where the Planck's constant is null, if it is to have an exact theory or, if it is not the case for an exact theory, that constant should be so small that in some conditions its square can be neglected.

This dual property of the potential viz. of cumulating both quantum and non-quantum properties at once should have been somehow embarrasing for physics, which always sought, mostly implicitly, by 'instinct' as it were, for ways to avoid it. For once, the reason can be made obvious if we assume a philosophically objective point of view: both the classical theory of hydrodynamics and the rule of quantization are incomplete as concepts, and cannot be taken as laws. The particulate structure of the continuum seems therefore hard to be accommodated theoretically, in order to justify the idea of potential for the Madelung fluid. In fact, all the issues raised above may add to nothing when compared to the fact that the definitions of the concepts related to 'quantum contributions' are purely formal, being obtained from one single function the density by *different mathematical operations*. Starting from equation (2.12) and going toward (2.15) everything comes out from the density by different applications of the operator ∇. In this respect, the equation (2.12) is actually an identity. This is not to deny the often-quoted idea that the mathematics is actually a tautology: the problem is not with the equations (2.16) and (2.17), but *with a physical content associated to these equations*.

On the other hand, however, as we only mentioned before, the fundamental rule of quantization is simply the consequence of an approximation of the image of classical atom, entailed by the fact that in the atomic phenomenology the electromagnetic fields play a substantial part in the observation of the microcosmic world. This model approximation, to wit, the harmonic oscillator model of hydrogen atom, serves the purpose of defining the measurement at microscopic level and by measurement to introduce the quantum rules. It was, indeed, the Thomas-Reiche-Kuhn rule the one that prompted Werner Heisenberg into his interpretation of the quantal laws of composition of quantum symbols (Heisenberg, 1925), and that rule was originally referring to an electric dipole oscillator model for atom [see (Mazilu & Porumbreanu, 2018) for details related to the rise of quantum mechanics], whereby the atom is electromagnetically described by a Hertz potential in two dimensions. It is therefore worth stepping into a case which involves the very rules of quantization according to Schrödinger view, in constructing the quantal quantities corresponding to some classical physical quantities (Wigner, 1954). Incidentally, as far as we are aware, this case seems quite singular in theoretical physics in fact it was intended just as an illustration of a 'would be' situation in the process of quantization but, taken as describing the reality of a Madelung fluid, it shows a neat difference between the two parts of the potential brought about by the Madelung's discovery.

Classical and Quantum Conservation Laws

The stochastic approach, essential to Nottale's scale relativity theory, brings to the fore an important issue related to the general description of the motion in physics: *the choice of initial conditions.* Its importance is not quite so obvious in classical mechanics, but the wave mechanics needs its due consideration. A first quotation from a classical work, resuming the theoretical physical knowledge contemporary with Madelung's article, is essential in understanding the point of view:

> In this connection one feature of the present processes perhaps deserves mention, because of its difference from what we are accustomed to in dynamics. There we think of *a particle describing a trajectory and can take any point of the trajectory indifferently as starting point of the motion.* But in the wave theory *the experimental conditions always mark out some special position,* say a slit, as starting point, and *at other places the waves will have spread.* Thus, unlike the case of dynamics, *we do not expect to get a solution in which the starting point is quite indifferent.* [(Darwin, 1927); *our Italics*]

The important work of Charles Galton Darwin from which we extracted this fragment, seems to us the best summary of the of the state of specific knowledge at the time of creation of the wave and quantum mechanics. We shall take the liberty of quoting it here from time to time, just to mark, in the best possible way we think, the differences between the classical approach of natural philosophy and the quantum mechanical approach.

To wit, in the above excerpt the emphasis is placed upon *experimental conditions* related to the idea of motion, which neatly differentiate the wave-mechanical approach of the motion, from its classical counterpart: "a particle describing a trajectory". The 'difference' introduced by the 'starting point' in the wave mechanical approach is instrumental in recognizing that a stochastic theory is needed within a SRT-type theory. This was subsequently recognized by Richard Feynman, who has built the archetype of any modern stochastic theory of wave mechanics [(Feynman, 1948); (Feynman & Hibbs, 1965)]. Reserving a later return to the content of the above excerpt, we now contemplate a first attitude in describing a possible connection between wave mechanics and classical mechanics: *the statistical connection* (Ehrenfest, 1927). This attitude regards only the concept of 'particle describing a trajectory', as it comes out from classical dynamics. The statistical connection known as Ehrenfest's theorem, then gives the classical quantities involved in the second law of Newton as means of the corresponding wave-mechanical ones, constructed on ensembles defined by probability densities as determined by the wave function.

The original Ehrenfest's theorem is therefore referring to the correlation between acceleration and force, as involved in the second law of Newton, and Eugene Wigner's work cited above can only incidentally be related to this subject matter. However, it is important both for the content of this section of our work and for the whole work in general, as we shall see, in connection with equation (2.14), for this equation addresses a subtle point which brings Wigner's idea in actuality. Namely, Wigner undertakes Ehrenfest's idea but for *a force proportional with the velocity and not with the acceleration*. And, by introducing the quantum contribution to potential, one has to recognize indeed the existence of a force proportional to velocity to be added as a quantum effect to that caused by classical inertia. Indeed, the equation (2.14) can be transcribed in Wigner's form, but for a single classical material point:

$$m\mathbf{u}(\mathbf{x}, t) = \nabla f(\mathbf{x}, t); \quad f(\mathbf{x}, t) = \frac{\hbar}{\sqrt{2}} \ln \rho(\mathbf{x}, t). \qquad (2.18)$$

From this perspective, Wigner's analysis and, more to the point, its very conclusions, carry a universal validity, pinpointing the true role of the statistical ensembles in the economy of a quantum theory in general. As a matter of fact, it is worth noticing again that the very substance of the Nottale's SRT is revealed by the stochastic aspects of the Schrödinger approach, based on the definition of a velocity field, as in equation (1.15). Such a definition is in turn a consequence of the Edward Nelson's stochastic approach of constructing the Schrödinger equation starting from a stochastic velocity field like that from equation (2.18). The velocity $\mathbf{u}(\mathbf{x}, t)$ is what, following the classical works of Einstein on Brownian motion, Nelson calls an *osmotic velocity field* (Nelson, 1966). To the extent to which the SRT needs a stochastic approach, a Wigner-type analysis should be essential here.

According to Wigner, the law of motion associated with those coordinates which have their rates given by the field velocity $\mathbf{u}(\mathbf{x}, t)$ in the form

$$m\dot{\mathbf{x}}(t) = \nabla f(\mathbf{x}, t) \qquad (2.19)$$

can be taken as expression of the simplest law of motion: *if no force acts, then the material point is at rest*. Now, if the potential $f(\mathbf{x}, t)$ is indeed referring to forces, then these forces should be of a special nature: the *dissipation forces*, proportional to velocity. The conservation of momentum is here equivalent with the statement that the center of mass is fixed for an isolated system of material points. And Eugene Wigner quantizes such a system by adopting the Ehrenfest's approach, in a theorem which can be rightfully considered counterpart of the original one, even though it was presented as merely mimicking it. The general description of the situation requires a nonstationary Schrödinger equation for a wave function characterizing a system of 'n' material points:

$$\frac{\partial}{\partial t} \Psi(\mathbf{x}_1, \mathbf{x}_2, ..., \mathbf{x}_n) = Q \Psi(\mathbf{x}_1, \mathbf{x}_2, ..., \mathbf{x}_n). \qquad (2.20)$$

Now, according to Ehrenfest's idea, the center of mass of such a system should be at rest. The coordinates of this center of mass are given by equations involving the probability density related to the wave function:

$$\overline{\mathbf{x}}_\alpha = \int \mathbf{x}_\alpha |\Psi|^2 \, d\tau, \quad m_\alpha \frac{d\overline{\mathbf{x}}_\alpha}{dt} = \int (\nabla_a f)|\Psi|^2 \, d\tau \qquad (2.21)$$

where $d\tau$ is the volume element of the configuration space of the system of those 'n' material points, and the integral extends over the whole configuration space. The second relation here is the equivalent of Ehrenfest's original theorem, which contains the second derivative of the mean position, instead of the first derivative appearing in the left hand side here. The requirement that the probability determined by the wave function should be independent of time, comes down to the the fact that Q in (2.20) should be anti-Hermitean. This leads to the commutation relations

$$[\mathbf{x}_\alpha, Q] = \frac{1}{m_\alpha}(\nabla_\alpha f) \qquad (2.22)$$

and further on to the most general anti-Hermitean form of the evolution operator:

$$Q = \sum_\alpha \frac{1}{m_\alpha}\left[(\nabla_\alpha f) \cdot \nabla_\alpha + \frac{1}{2}\nabla_\alpha^2 f \right] + ig \qquad (2.23)$$

where 'g' is invariant with respect to displacements and rotations, otherwise arbitrary. Thus, we have the energy equation

$$E = i\hbar \int \Psi^* \left\{ \sum_\alpha \frac{1}{m_\alpha}\left[(\nabla_\alpha f) \cdot \nabla_\alpha \Psi + \frac{1}{2}\Psi\nabla_\alpha^2 f \right] + ig\Psi \right\} d\tau \qquad (2.24)$$

the angular momentum conservation, which is the same with that of the usual wave mechanics:

$$\mathbf{M} = i\hbar \int \Psi^* \left\{ \sum_\alpha (\mathbf{x}_\alpha \times \nabla_\alpha)\Psi \right\} d\tau \qquad (2.25)$$

and the very momentum conservation, to which we have to add the center of mass conservation laws, analogous to classical ones, viz.:

$$\mathbf{P} = -i\hbar \int \Psi^* \left(\sum_\alpha \nabla_\alpha \Psi \right) d\tau;$$

$$\mathbf{R} = \int \left(\sum_\alpha m_\alpha \mathbf{x}_\alpha \right)(\Psi^* \Psi) d\tau. \qquad (2.26)$$

Whence the Wigner's conclusions:

... The equations of motion (2.18) *would have made* the physics of the past fifty years very much easier : they would have made it *impossible to introduce the theory of relativity*, and quantization of the equations *would not have changed* their physical content. The only new feature which the quantum theory introduces is the *complex phase of our wave function* Ψ and *is questionable whether this quantity could be attributed any physical significance*. Since the quantum conservation laws (2.24), (2.25) and (2.26) *are all based on this complex phase*, and vanish for a real wave function Ψ, *their physical interpretation is open to question*.

The above example was meant as a warning against a facile identification of symmetry and conservation laws. It reminds us that the Hamiltonian formulation is necessary for that connection to hold in ordinary mechanics and that *while it is always possible in quantum theory to deduce conservation laws from a symmetry* condition, the *interpretation of these conservation laws*, and their significance, *might be quite problematical.*

The fact that the quantization of the equations of motion would not lead to a real quantum theory might have been foreseen from the fact that the uncertainty principle is hardly compatible with (2.18). If all the coordinates have sharp values, this holds also for *the forces* $-$ grad(f). This is true also in current quantum theory. However, in the present theory, *the forces determine the velocities*, rather than the accelerations, and these become determined also. [(Wigner, 1954); *our Italics*]

The Madelung fluid is one case illustrating how "the forces detemine the velocities", however at a different level of statistics, as we shall see later. This statistics is related to the "complex phase of the wave function" whose physical interpretation turned out a half a century later. However, for the moment being, this interpretation forces us to an observation about the potential, that appears by and large to have been unnoticed thus far.

In order to get the ideas in their proper order, we need to take notice of the fact that in his attempt of *physical interpretation*, Erwin Madelung, like Schrödinger himself, seeks actually for a physical explanation of *the atom as a physical structure represented by the classical planetary model*, for this is the actual physical image of the atom. This seems somehow contrary to the quantum mechanics which seeks only for the description of a model atom befitting the idea of measurement. Then, we here too should rather pursue a way in which both the hydrodynamics and the quantization refer to models of things, not to models approximating these things. So let us turn back to assuming the Schrödinger equation, the way Madelung assumed it originally, i.e. without considering that the density could be somehow referred to as a probability density. This time, therefore, we consider fluid dynamics without resorting to the Max Born's type of interpretation of the wave function, or any interpretation of the kind, for that matter.

Hydrodynamics of Free Point Particles: Universality of the Schrödinger Equation

The "complex phase" of the wave function is, even to Schrödinger himself, the mark of the solution of nonstationary equation. However, his justification for the existence of the essentially complex wave function can be recognized only as anecdotal [(Schrödinger, 1933), Footnote 1, p. 166; see also (Mazilu & Porumbreanu, 2018)]. Such a solution is legal though, and beyond any possibility of rejection, so to speak, provided some extra constraints are applied to the classical mathematics involved in this problem. For instance, if the nonstationary Schrödinger equation proves to be a necessary tool of the trade, then the classical properties of the complex phase ensue just naturally by the way of mathematics. Indeed, the essentially complex Schrödinger equation leading to previous results is:

$$
\begin{aligned}
&A\Big(\frac{\partial \Phi}{\partial t} + \frac{1}{2m}(\nabla\Phi)^2 + V\Big) \\
&= i\hbar\Big(\frac{\partial A}{\partial t} + \frac{A}{2m}\nabla^2\Phi + \frac{1}{m}(\nabla A)\cdot(\nabla\Phi)\Big) + \frac{\hbar^2}{2m}\nabla^2 A
\end{aligned}
\tag{2.27}
$$

where only the first relation from equation (2.4) was assumed, i.e. a wave function of the regular complex form in terms involving the amplitude and phase:

$$
\Psi(\mathbf{x}, t) \quad A(\mathbf{x}, t)e^{i\Phi(\mathbf{x}, t)}
\tag{2.28}
$$

without any further assumptions about their meaning within the framework of the classical physics. That is, none other than those assumptions necessary for the mathematical handling toward some results, of course. Now, in order to get those results, the equation (2.27) must be read somehow, and this can be done in many different ways, each one of them leading to just as many different results. The Madelung's original way of reading was based on the *algebra of complex numbers*, whereby a number is zero only if its two real components are both zero concurrently. However if, in view of its primary importance here, *the physical way of reading* mathematical equations is to prevail, the algebraic structure should be only a subordinate one, and in this case the equation (2.27) has a clear advantage over (2.17). This last equation shows that the *classical action can be the phase of wave function regardless its amplitude*, only if the square of the Planck's constant is negligible. That might not be the most general case, as equation (2.27) shows. Indeed, in this last equation the quantum terms are clearly exhibited within two orders of magnitude in the Planck's constant, appearing in the right hand side. If the complex algebra is to be subordinate to a 'physical algebra', so to speak, then equation (2.27) must be read as follows: the exponent of wave function (2.28) is the classical action if, and only if, its amplitude is a nontrivial solution of the Laplace equation, with its square satisfying a continuity equation. Indeed, we have

$$\frac{\partial \Phi}{\partial t} + \frac{1}{2m}(\nabla\Phi)^2 + V = 0 \quad \Leftrightarrow \quad \begin{aligned} &\nabla^2 A = 0; \\ &\frac{\partial A^2}{\partial t} + \frac{1}{m}\nabla \cdot [A^2(\nabla\Phi)] = 0, \end{aligned} \tag{2.29}$$

which represents the whole logic just laid down above. The mark of classical action is here taken to be the classical Hamilton-Jacobi equation which, once satisfied by the phase $\Phi(\mathbf{x},t)$, means the vanishing of the left hand side of equation (2.27) and, of course, vice versa. Then the right hand side of equation (2.27) is vanishing but, as a complex quantity now, it is vanishing only when its real components are both vanishing concurrently, as in the Madelung's case. This gives the right hand side in equation (2.29). Note, nevertheless, that now *it is not the phase that should not be zero, but the amplitude* of the wave function. This fact dramatically changes the emphasis, with very important consequences, as we shall see later.

This approach, however, shows that the scheme of thinking thus followed would imply that, if the wave function is considered as a general representation of a physical situation, then the classical description (the Hamilton-Jacobi equation) inflicts specific restrictions upon the very amplitude of the signal represented by that wave function. The most important of these restrictions is that *the amplitude of such signal has to be a solution of the Laplace equation*. These restrictions might not be quite unnatural, but they are altogether independent of the value of the Planck's constant, and if we do not know anything about the Schrödinger equation, we might very well dispense with it. So, the situation raises an obvious question: *is the nonstationary Schrödinger equation necessary to our knowledge?* And while we are at it, let us take notice of the fact that nothing has been said here about the *stationary Schrödinger equation*. It may be a consequence of the nonstationary corresponding equation, which is true indeed, but not in a historical and logical order of things. For instance, the Madelung's approach follows the Schrödinger's approach, which starts with the *stationary equation*, and then builds the nonstationary one by eliminating the energy with the help of the first time derivative of the wave function.

Now if, for the sake of internal completeness of the theory say, we hold the existence of a stationary equation as one essential ingredient of that theory, this feature is not at all missing from among the possibilities offered to us by equation (2.27). Consider, indeed, the Madelung fluid as a *fluid of free particles as required by the scale relativity theory of Laurent Nottale*. We even have here a quintessential physical example: the classical ideal gas which served, among other as the basis of modern thermodynamics. Then we can also read the equation (2.27) as follows:

$$\frac{\partial \Phi}{\partial t} + \frac{1}{2m}(\nabla\Phi)^2 = 0; \quad \frac{\partial A^2}{\partial t} + \frac{1}{m}\nabla \cdot [A^2(\nabla\Phi)] = 0 \qquad (2.30)$$

provided the amplitude $A(\mathbf{x}, t)$ of the wave function from (2.28) is a solution of the *stationary Schrödinger equation*:

$$\frac{\hbar^2}{2m}\nabla^2 A = V(\mathbf{x}, t)A. \qquad (2.31)$$

The two equations in (2.30) are obviously independent of *any* potential. One can hope to solve for the action function $\Phi(\mathbf{x}, t)$, as this operation was done for long in the classical mechanics, and then, based on this solution, to find a corresponding amplitude, whose square should be a density, as the second equation from (2.30) shows it. However, in this case, the equation (2.31) offers just a definition for the potential, which thereby acquires quite a natural feature: *it is purely quantal*. If it is zero, from (2.31) we have (2.29) as a special case. However, in this reading the potential seems to be redundant, if there is no need for the quantum as represented by the (square of) Planck's constant. So, this time there is no way around: we need to prove the necessity of the quantum theory itself and, moreover, that it *is naturally connected with the idea of potential, not with that of wave function*.

Strange as it may seem, this approach is the only one satisfying the condition noticed by us beforehand, that the quantization is related to an approximate model of the atom: the electromagnetic Hertz dipole. For, then, the fundamental quantum rule can be surely recognized as an expression of electromagnetic interaction, as it was indeed the case, concurrently with the Planck's constant (Boyer, 1969, 1975). Therefore the presence of a quantum should, indeed, be recognized only in the expression of potential, while the wave function must contain even some other kinds of physical magnitudes. This brings to fore the important problem of connection between the stationary and nonstationary Schrödinger equation, in case this last one is altogether necessary to our knowledge. Further, it asks for a precise definition of what is known as *physical interpretation*, and of interpretation in general, for that matter.

A Definition of the Interpretation

The scheme contained in the equations (2.30) and (2.31) seems to be rather simple: the general signal (2.28) describes a 'swarm' of free classical particles of a continuum of density A^2 , each one of them having a momentum $m\mathbf{v} = \nabla\Phi$. Notice then that the whole function $\Psi(\mathbf{x}, t)$ can be recovered, in principle, from equation (2.30). In this case the equation (2.31) is nothing else, indeed, but a definition of the potential. Or else, knowing the potential we can solve (2.31) for the amplitude, and then find a solution of the first equation (2.30), compatible with this solution via the second equation from (2.30). If, for instance, we are trying to give an explanation to that "infinite free path" mentioned by Erwin Madelung, we may have no other choice but to notice that it begs for a kind of generalization of the modern principle of asymptotic freedom: *in a region of pure quantum forces the wave function describes an ensemble of free particles.* For, indeed, the stationary Schrödinger equation (2.31) shows that the continuum described by the function (2.28), as structured by a swarm of free particles in the manner proposed by Madelung, appears as an ensemble of particles evolving under a purely quantal potential: *the whole potential,* not just part of it, is in fact a "quantum contribution". Thus, the description of the continuum by the function (2.28) is a purely undulatory description of a region of space-time. Its interior however, is a swarm of free classical particles described by an equation of continuity for a density proportional to the square of the modulus of function (2.28). Each one of these classical particles has a momentum given by the gradient of phase of (2.28). However, a question still stands: *what this very region represents?* The answer to this question cannot be given classically, for the classical argument has already been exhausted. But it can be given mathematically, provided we are free to choose *an interpretation* of the scheme contained in the equations (2.30).

Fact is that any signal represented in the complex form (2.28) is the solution of an equation resembling the free particle Schrödinger equation, provided some specific conditions are in effect (Schleich, Greenberger, Kobe & Scully, 2013). The complex form we are talking about assumes *an implicit time and space dependence* for the phase and the amplitude of the signal (2.28), resulting in *the identity:*

$$i\frac{\partial\Psi(\mathbf{x}, t)}{\partial t} + \beta\nabla^2\Psi(\mathbf{x}, t)$$

$$= -\left(\frac{\partial\Phi(\mathbf{x}, t)}{\partial t} + \beta[\nabla\Phi(\mathbf{x}, t)]^2 - \beta\frac{\nabla^2 A(\mathbf{x}, t)}{A(\mathbf{x}, t)}\right) \qquad (2.32)$$

$$+ \frac{i}{2A^2}\left(\frac{\partial A^2(\mathbf{x}, t)}{\partial t} + 2\beta\nabla\cdot[A^2\nabla\Phi(\mathbf{x}, t)]\right),$$

where β is a constant having the physical dimensions of a rate of area (m^2/s). The 'specific conditions' necessary in order that $\Psi(\mathbf{x}, t)$ be a solution of the time dependent Schrödinger-type equation for the free particle, to wit:

$$i\frac{\partial \Psi(\mathbf{x}, t)}{\partial t} + \beta \nabla^2 \Psi(\mathbf{x}, t) = 0 \tag{2.33}$$

are in fact the two known equations that guarantee the vanishing of the right hand side of equation (2.32), so that (2.33) can take place:

$$\frac{\partial \Phi(\mathbf{x}, t)}{\partial t} + \beta [\nabla \Phi(\mathbf{x}, t)]^2 - \beta \frac{\nabla^2 A(\mathbf{x}, t)}{A(\mathbf{x}, t)} = 0,$$
$$\frac{\partial A^2(\mathbf{x}, t)}{\partial t} + 2\beta \nabla \cdot [A^2 \nabla \Phi(\mathbf{x}, t)] = 0. \tag{2.34}$$

The first of these equations is the classical Hamilton-Jacobi equation, the second one is a continuity equation for a fluid of density given by the square of the amplitude of the function Ψ. Here, however, by comparison with the equation (2.29), *the classical potential* suggested by the Hamilton-Jacoby equation for the phase of function Ψ is only defined by the amplitude of this function, through equation

$$V(\mathbf{x}, t) \equiv -\beta \frac{\nabla^2 A(\mathbf{x}, t)}{A(\mathbf{x}, t)} \tag{2.35}$$

which can be turned into a stationary Schrödinger equation. In other words, the nonstationary Schrödinger equation for the free particles is a mathematical tool of unquestionable existence, just like any other mathematical concept used by physics, as long as the wave function has to be complex. The problems rise with the equation (2.35), and they concern only the mathematical structure of the functions representing the amplitude and the phase of the wave function. These should tell us what is a free material particle from the point of view of wave mechanics. When it comes to considering the potential, the phase of the wave function cannot be identified with the classical action quite unconditionally: this last one must satisfy the above constraints. Only in this instance can one have the freedom of using the potential in a two-way reasoning: either as given and then helping to find the amplitude or, once the amplitude known, it helps to find the potential. With a proper extension of the concept of complex number, this last way is the old one initiated in the physics of elementary particles by Tullio Regge (Regge, 1959).

Thus, we can say that the equation (2.33) is indeed a universal instrument of our knowledge, once it ensues mathematically from a necessary complex form of the wave function. Then everything comes down to the *interpretation of the wave function*, which should be part of a *general interpretation process*, and this is the moment where the idea of ensemble makes its proper entrance in the argument. Like in all of the classical cases, the ensemble enters first by its historical element — the particle — as in the quintessential case of classical ideal gas. We return again to C. G. Darwin for a brilliant choice of the proper words characterizing the physical situation:

> It is almost impossible *to describe the result of any exper-iment except in terms of particles* — a scintillation, a deposit on a plate, etc. — and this language is quite *incompatible with the language of waves*, which is used in the solution. A nec-essary part of the discussion of any problem is therefore *the translation of the formal mathematical solution, which is in wave form, into terms of particles. We shall call this process the interpretation*, and only use the word in this technical sense [(Darwin, 1927); *our Italics*]

Notice, in this context, that the equation (2.33) is the Schrödinger nonstation-ary equation for the free particles, once no potential is involved in it. However, classically, these particles are not free, as the first of the equations (2.34) shows. Consequently, if the function $\Psi(\mathbf{x}, t)$ itself represents an *ensemble of free parti-cles* as required for a proper physical interpretation, these are free particles *not from classical point of view, but from the Schrödinger equation point of view:* classically they can be anything along the line of physical freedom.

In order to represent the classical case of free particle, the amplitude of the wave function should be taken, for instance, from among the solutions of the partial differential equation (2.35), with the potential $V(\mathbf{x}, t) \equiv constant$. According to classical dynamical standards, such a potential would mean null force at the positions where it has that constant value, therefore free particles as the material particles upon which the force acts. Let us find such solutions: designate the constant, say with V_0 ; then the equation (2.35) assumes the form of a Helmholtz equation

$$\nabla^2 A(\mathbf{x}, t) \pm \sigma^2 A(\mathbf{x}, t) = 0, \qquad (2.36)$$

with $\sigma^2 \equiv |V_0/\beta|$. The notation was chosen in order to suggest that the potential may assume any constant value, positive or negative. It may have even a complex value, for that matter, but just for the sake of general argument we do not go momentarily that far. A separation of variables via identity $A(\mathbf{x}, t) \equiv R(x, t) \cdot F(\theta, \phi, t)$, where $x \equiv |\mathbf{x}|$ and (θ, ϕ) are spherical polar angles, helps bringing (2.36), for the case of minus sign, to the form

$$\frac{1}{R}(x^2 R')' - \sigma^2 x^2 = -\frac{\Delta_{\theta\varphi} F}{F} = k(k+1). \qquad (2.37)$$

Here a prime means derivative with respect to the unique variable as usual, and 'k' is an integer, accounting for the periodicity of solution on the unit sphere.

Thus the radial part of the amplitude of this wave function should be a solution of the equation

$$x^2 R'' + 2x R' - [k(k+1) + \sigma^2 x^2] R = 0 \qquad (2.38)$$

while the angular part is a regular spherical harmonic. This last equation is a transform of the standard Bessel equation given by Frank Bowman [(Bowman, 1958); see his equation (6.80)], and has the general solution of the form:

$$R_k(x,t) = C_+(t) \frac{I_{k+1/2}(\sigma x)}{\sqrt{x}} + C_-(t) \frac{I_{-(k+1/2)}}{\sqrt{x}}(\sigma x) \qquad (2.39)$$

where I_n are modified Bessel functions of the first kind. Likewise, for the cases with a plus sign in equation (2.36), the solution is:

$$R_k(x,t) = C_+(t) \frac{J_{k+1/2}(\sigma x)}{\sqrt{x}} + C_-(t) \frac{J_{-(k+1/2)}}{\sqrt{x}}(\sigma x) \qquad (2.40)$$

where J_n are the usual Bessel function of the first kind.

According to the philosophy delineated by C. G. Darwin in the excerpt right above, in order to be considered wave functions, the solutions (2.39) and (2.40) *need interpretation*. We are now in position to state that this interpretation is related to the SL(2,R)-type algebraical structures, and will be revealed in its details further in the present work. For now, it suffices to disclose the key of this interpretation, which resides is in the form of the radial part of these solutions. Both the classical mechanics and the general relativity contain a clear possibility of such an interpretation for the case of the so called free fall in a gravitational field. It thus becomes obvious that we need to put this interpretation under the concepts related to the nonstationary Schrödinger equation for the free particle, insofar as this equation is a fundamental mathematical instrument.

Fact is that the nonstationary Schrödinger equation admits, besides the clasical Galilei group proper, an extra set of symmetries (Niederer, 1972) that, in general conditions, can be taken in a form involving just one space dimension and time, as a SL(2,R) type group in two variables with three parameters (de Alfaro, Fubini & Furlan, 1976). Limiting the general conditions, the space dimension can be chosen as the radial coordinate in a free fall, as in the case of Galilei kinematics, which can also be extended *as such* in general relativity (Herrero & Morales, 1999, 2010), for instance in the case of free fall in a Schwarzschild field. The essentials of the argument of Alicia Herrero's and Juan Antonio Morales' work just cited are delineated based on the fact that the radial motion in a Minkowski spacetime should be a conformal Killing field, which is a three-parameter realization of the $\mathbf{sl}(2, R)$ algebra in time and the radial coordinate. This is a Riemannian manifold of the Bianchi type VIII (or even type IX, forcing the concepts a little) when taking the stand of one of the epoch-making, widespread, nomenclatures of the theory of general relativity [(Bianchi, 2001); see especially the editorial comment for this English version of the classical work of Luigi Bianchi]. The bottom line here is that, as long as the general relativity is involved, the nonstationary Schrödinger equation describes the continuity of matter. And since, as a universal instrument of knowledge, this Schrödinger equation is referring to free particles, we need to show what kind of freedom is this in classical terms.

For our current necessities it is best to start with the finite equations of the specific SL(2,R) group, and build gradually upon these (Mazilu & Porumbreanu, 2018), in order to discover the connotations we are seeking for. Working in the variables (t, x) as above, the finite equations of this group are given by the transformations:

$$t \to \frac{\alpha t + \beta}{\gamma t + \delta}; \quad x \to \frac{x}{\gamma t + \delta}. \tag{2.41}$$

This transformation is a realization of the SL(2,R) structure in variables (t,x), with three essential parameters (one of the four constants α, β, γ and δ is superfluous here). Every vector in the tangent space SL(2,R) is a linear combination of three fundamental vectors, the infinitesimal generators:

$$\mathbf{X}_1 = \frac{\partial}{\partial t}, \quad \mathbf{X}_2 = t\frac{\partial}{\partial t} + \frac{x}{2}\frac{\partial}{\partial x}, \quad \mathbf{X}_3 = t^2\frac{\partial}{\partial t} + tx\frac{\partial}{\partial x} \tag{2.42}$$

satisfying the basic structure equations:

$$[\mathbf{X}_1, \mathbf{X}_2] = \mathbf{X}_1, \quad [\mathbf{X}_2, \mathbf{X}_3] = \mathbf{X}_3, \quad [\mathbf{X}_3, \mathbf{X}_1] = -2\mathbf{X}_2 \tag{2.43}$$

which we take as standard commutation relations for this type of algebraic structure, all along the present work. The exponential group has an invariant function, which can be obtained as the solution of a partial differential equation:

$$(c\mathbf{X}_1 + 2b\mathbf{X}_2 + a\mathbf{X}_3)f(t,x) = 0 \therefore$$
$$(at^2 + 2bt + c)\frac{\partial f}{\partial t} + (at + b)x\frac{\partial f}{\partial x} = 0. \tag{2.44}$$

The general solution of this equation is a function of the constant values of the ratio:

$$\frac{x^2}{at^2 + 2bt + c} \tag{2.45}$$

which represents the different paths of transitivity of the action described by (2.42).

In order to draw some proper conclusions from these mathematical facts, let us go back to the transformation (2.41) and consider it from the point of view of classical physics. The first principle of dynamics offers a special content to the classical time: it is represented by the uniform motion of a free classical material particle. According to this principle, such a particle is free as long as no forces act upon it. And the equation (2.45) faithfully records this idea in an obvious form: the paths of transitivity of the action (2.41) are given by the 'radial motion' of a free classical material point, no question about that. Questions rise, however, and on multiple levels at that, when noticing that the general solution of equation (2.44) is an *arbitrary function* of the ratio (2.45). For once, we are compelled to notice that the *content of time* in (2.41) is not classical anymore, at least not in general, being a ratio of coordinates representing *two uniform motions*. Likewise, the second equation (2.41) can be taken as representing the *content of spatial coordinate* of the motion in terms of the classical coordinate of a uniform motion. This much, at least, can be put in the common charge of the wave mechanics and general relativity, regarding an 'updating' of the idea of time and space contents. But there is more to it, regarding the concept of freedom, because at this point we start to notice some apparently unrelated facts from the past, which seem to pick up concrete shapes, all converging to the ratio from equation (2.45).

First, comes the second of Kepler laws, viz. that law serving to Newton as a means to introduce the idea of a center of force: if, with respect to such a material point, a motion proceeds according to the second Kepler law, then the field of force should be Newtonian. The wave mechanics shows that this law means more than it was intended for initially, namely that it should have a statistical meaning, according to the idea of Planck's quantization (Mazilu & Porumbreanu, 2018). Indeed, if 'x' denotes the distance of the moving material point from the center of force,

$$x^2 d\theta = \dot{a}dt \quad \therefore \quad \dot{a}\frac{dt}{d\theta} = x^2 \qquad (2.46)$$

where θ is the central angle of the position vector of the moving material point with respect to the center of force. In this form the law usually serves as a transformation in the mathematical treatment the central motion. However, from the point of view of the physical content of time, the second equality in equation (2.46) tells us much more if we take the argument out of the mathematical form of the classical Kepler problem.

To wit, consider an *extended body* revolving in a central field of Newtonian forces. It can be imagined as a swarm of classical material points, and such a swarm illustrates classical laws, provided it is considered as a swarm of free material points in the classical sense of the word (Larmor, 1900). In the first of equations (2.41) this requirement would mean that the material points are considered simultaneously. Then each material point can be located in the swarm by four homogeneous coordinates $(\alpha, \beta, \gamma, \delta)$, or three nonhomogeneous coordinates, if the equations (2.41) represent the content of time and radial coordinate for the space region covered by this body. The simultaneity in the motion of the swarm of material points can be differentially characterized, giving a Riccati equation in pure differentials:

$$d\frac{\alpha t + \beta}{\gamma t + \delta} = 0 \quad \therefore \quad dt = \omega^1 t^2 + \omega^2 t + \omega^3. \tag{2.47}$$

Thus, for the description of the extended body in motion as a succession of states of the ensemble of simultaneous material points, it suffices to have three differential forms, representing a coframe of the **sl**(2,R) algebra:

$$\omega^1 = \frac{\alpha d\gamma - \gamma d\alpha}{\alpha\delta - \beta\gamma}; \quad \omega^2 = \frac{\alpha d\delta - \delta d\alpha + \beta d\gamma - \gamma d\beta}{\alpha\delta - \beta\gamma};$$

$$\omega^3 = \frac{\beta d\delta - \delta d\beta}{\alpha\delta - \beta\gamma}. \tag{2.48}$$

That this coframe refers to such an algebra, can be checked by direct calculation of the Maurer- Cartan equations which are characteristic to this algebra:

$$d \wedge \omega^1 - \omega^1 \wedge \omega^2 = 0;$$

$$d \wedge \omega^2 + 2(\omega^3 \wedge \omega^1) = 0; \tag{2.49}$$

$$d \wedge \omega^3 - \omega^2 \wedge \omega^3 = 0.$$

Using these conditions one can prove that the right hand side of equation (2.47) is an exact differential (Cartan, 1951), therefore it should always have an integral. The Cartan-Killing metric of this coframe is given by the quadratic form $(\omega^2/2)^2 - \omega^1\omega^3$, so that a state of an extended orbiting body in the Kepler motion, can be organized as a metric phase space, a Riemannian three-dimensional space at that. The geodesics of this Riemannian space, are given by some conservation laws of equations

$$\omega^1 = a^1(d\theta); \quad \omega^2 = 2a^2(d\theta); \quad \omega^3 = a^3(d\theta) \tag{2.50}$$

where $a^{1,2,3}$ are constants and θ is the affine parameter of the geodesics, so that, along these geodesics the differential equation (2.47) is an ordinary differential equation of Riccati type:

$$\frac{dt}{d\theta} = a^1 t^2 + 2a^2 t + a^3. \tag{2.51}$$

This equation can be identified with (2.46), provided its right hand side is proportional to the square of a 'radial coordinate' of a free classical material point. Mathematically this requires an ensemble generated by a harmonic mapping between the positions in space and the material points, with the square of the radial coordinate 'x' measuring the variance characterizing the distribution of material points in space.

So everything comes down to finding a physical system whose motion gives the content of time by the first relation (2.41), the same way the uniform motion gives the content of time in the classical case. For this we need to treat the wave function (2.28) as a signal recorded in a given space point $\mathbf{x}$. In this case both the amplitude A and the phase Φ are to be considered as functions of time only. The experimental practice asks here for an analysis of signal in the time domain, allowing us to assign mechanical properties to the magnitudes extracted from the records. A mandatory parameter in this analysis is the *instantaneous frequency* (Mandel, 1974), which can be evaluated as the first time derivative of the phase of wave function. Denoting q(t) the wave function in this instance, i.e. as the instantaneous 'elongation' representing the recorded signal, we want to associate this signal with a mechanical oscillator, in order to have a physical interpretation of the parameters, especially of the phase as function of time. This association comes down to the following equivalences, representing connections between amplitude and phase as functions of time at a certain position:

$$\ddot{q}(t) + 2\lambda\dot{q}(t) + \omega_0^2 q = 0 \leftrightarrow \begin{array}{l} \dfrac{\ddot{A}}{A} + 2\lambda\dfrac{\dot{A}}{A} + \omega_0^2 = \dot{\Phi}^2; \\[2mm] \dfrac{\dot{A}}{A} + \dfrac{\dot{\Phi}}{2\phi} + \lambda = 0, \end{array} \qquad (2.52)$$

from which we have, denoting by $\{^*,^*\}$ the *Schwarzian derivative* of the first symbol in curly brackets with respect to the second one, we have:

$$\dot{\Phi}^2 + (1/2)\{\Phi, t\} = \omega_0^2 - \lambda^2. \qquad (2.53)$$

The signal having the instantaneous frequency as an exact mechanical frequency of a damped harmonic oscillator, should have $\{\Phi, t\} = 0$, which means

$$\Phi(t) = \frac{\alpha t + \beta}{\gamma t + \delta} \quad \therefore \quad \dot{\Phi} = \frac{\alpha\delta - \beta\gamma}{(\gamma t + \delta)^2}. \qquad (2.54)$$

Therefore the most general signal having mechanically defined parameters is of the form

$$q(t) = (at + b)e^{-\lambda t}exp\left(i\frac{\alpha t + \beta}{\gamma t + \delta}\right) \qquad (2.55)$$

and the time of the first equation (2.41) is the phase of such a signal, whose amplitude is a 'damped' uniform motion. Notice that for a proper choice of the arbitrary constants of integration 'a' and 'b', the equation (2.55) is, in fact, a special connection between the group variables from equation (2.41). Let us recount the results here, to better realize what we acquired.

In order to get a sharply defined instantaneous frequency of a signal by mechanical means, we need to compare this signal with a local harmonic oscillator. Then, the phase of this harmonic oscillator has to be a homographic function of the time moments involved in the time sequence defining the signal by its amplitudes. In this case, the instantaneous frequency of the signal has to be the reciprocal square of a linear function in the moments of this time sequence. In this case, by equation (2.55) the product between the square of instantaneous amplitude and the frequency of the signal must be a special function of the phase, to wit, a complex exponential. We need to insist here on the idea of *time content*, with a first observation which hints on the role of a relativity-like theory.

Obviously, when we speak of the content of time, we understand a time sequence: a set of moments of time provided arbitrarily, e.g. by a local clock, and ordered with the assistance of a local motion. Typically, such a motion is taken as the classical uniform motion, to be somehow locally gauged. Until further elaboration on the notion of 'local' here, we have to take it in the strict mathematical sense: in a given point in space. It remains, of course, to explain how the 'uniform motion' matches the idea of a space position, but for now let us assume that this is the case, and only notice how relativity bypassed the idea of 'local'. There are two aspects of the physical idea of time, which need to be somehow corroborated with one another in order to give a useful time concept. First we have the idea of time provided by a clock: a device measuring the time according to local needs, usually by a periodic process. Then we have the idea of choosing from this time *an ordered sequence*, and the 'device' for this ordering is a uniform motion. Classically, for a uniform motion, the time is a *parameter of continuity*. Physically, the problem occurred as to the sound definition of the ordered sequence of time itself, and this was the point where the special relativity started to erect its structures.

Indeed, the special relativity circumvented the necessity of a 'local gauging' of the uniform motion by the practical observation that we have always at our disposal an apparently 'global gauging', offered by the light phenomena. The physics then discovered that these phenomena can be locally described as electromagnetic phenomena. The Maxwellian form of electrodynamics shows that, as long as the time is classically considered i.e. as just a parameter of continuity this concept of local does not require any further consideration beyond in a given space position. At this point Albert Einstein stepped in (Einstein, 1905) with the observation that the ordered time sequences in different position in space can be set in correspondence by the procedure of synchronization. This requires the light as essential in building a special correspondence involving the

40

positions in space and the time moments, incorporated as such into the concept of spacetime. Further, Einstein noticed that the electromagnetic phenomena, as described by Maxwell equations, behave controllably with respect to the space-time correspondences thus conceived, an observation that encouraged the idea that light phenomena should be exclusively of an electromagnetic nature. This conclusion pushed, so to speak, in the background, for a while, the necessity of 'local gauging', but mixed inadvertently the two differentias of time. For, as long as Maxwell equations are involved, the time can be used indifferently, both as a continuity parameter, and as a time sequence. This was the state of the case until electrodynamics popped up again, with the concept of a Yang- Mills field, brought about by wave-mechanical necessities of physics (Yang & Mills, 1954).

This last concept did not come out of nowhere: it was reached via another route left unexplored by Einstein himself, but noticed by Louis de Broglie. It is the notion of energy content of a massive particle, according to which to any particle at rest a frequency can be associated, which behaves according to the rules of special relativity (de Broglie, 1923). This association came up with the *idea of quantum*, which, in turn, sprung from a special statistics (Mazilu, 2010) inspired and supported exclusively by the very electromagnetic image of the light [(Einstein, 1909, 1965); see also (Boyer, 1969, 1975)]. However, the initial de Broglie's theory ignored the statistical aspect of the concept of quantum, a fact which had the immediate consequence of forcing upon theory the notion of *wave group*. In the long term, however, this fact obligated de Broglie to recognize the necessity of a statistics for the very 'wave phenomenon called classical material point', after his own expression, and this statistics, referring this time to the rest mass, is exactly of the general type involved in the Planck's theory of radiation: a statistics described by an exponential probability distribution (de Broglie, 1966).

Coming back to our idea of the mechanical measurement of phase, we need to notice that it provides a method of measuring the phase, by gauging it with a harmonic oscillator. This method is, actually, a long known method of measurement in physical and technical problems involving periodical phenomena. And Louis de Broglie's idea of introducing the frequency for massive particles the same way as for light particles, gains thereby a special meaning: that of deciding what is 'finite', when having at our disposal just the notions of 'local' and the 'global', as introduced by special relativity. Insofar as the quantum is involves in defining the de Broglie frequency of a massive particle, the decision is clear: this 'finite' should be spatially *a cavity* like the one built by Wien and Lummer for the verification of the laws of thermal radiation (Wien and Lummer, 1895), wherein, however, *only the phase can be measured*. The observation was brought about by Lachlan Mackinnon, who thus showed how the idea of randomness is to be taken in the wave physics of Louis de Broglie (Mackinnon,

41

1978). Quoting:

> An observer is sitting in an *empty laboratory* in which there is one particle – say, an electron – stationary with respect to him. Suppose that *his powers of observation are restricted to observing the phase of the de Broglie wave* of this electron at any point in the laboratory. All that he will then be able to observe is that *the phase is uniform throughout the laboratory* and he *will not be able to use it to locate the electron*. Another observer enters the laboratory and walks smartly through it; he finds the phase of the electron's de Broglie wave far from uniform; he *tells the first observer what he sees*; the first observer then deduces that the electron *can only be at one of those points in the laboratory where he and the second observer agree about the phase*. Further observers enter the laboratory, *each moving with a different speed and in a different direction*, so that they each observe *a different de Broglie wave from the electron*; with their *additional information*, the first observer will deduce that the electron *can only be at the point or points where all these waves are in phase*. A sufficient randomness among the observers will ensure that *the electron can be located* because *the shared information* will have allowed the first observer to construct his de Broglie wave packet for the electron. The "anchor" for the *time coordinates of the reference frames of the various observers is also at the particle*, but, as will be pointed out in the next paragraph but one, the apparent frequencies of all the de Broglie waves are the same to the first observer, so the wave packet can be constructed *from the space information alone.* [(Mackinnon, 1978); *our Italics*]

Now, in view of our results right above, this excerpt gains significance beyond the special theory of relativity. Indeed if the measurement of phase is done as usual, then "observing the phase" means the possibility of construction of a phase as in equation (2.54), based on which we define an instantaneous frequency. The 'uniform phase throughout laboratory' then comes down to equation (2.47), which means that the cavity is physically a Riemannian space given by the **sl**(2,R) algebra, whose coframe is given in the equation (2.48). The 'agreement about the phase' is then a statistical matter to be decided as in equation (2.51), which gives any sequence of time moments as means of an ensemble characterized by an exponential distribution with quadratic variance function, which is a Planck-type ensemble that led to the first quantization (Mazilu, 2010).

This may be a first incentive in taking the nonstationary Schrödinger equation as fundamental to physical thinking, forasmuch as the particles in such a cavity should be free particles. But then, the whole quantum philosophy is concentrated on the potential, as we just said, and we need to find a way to bring together the potential and the wave function. We shall do this here, by following logically the historical order of things, in explaining the idea of an 'empty laboratory' in transition between scales, as *a fundamental concept of scale relativity*. This concept allows us to build a model of space extended particle, by *interpretation* of some well known physical cases, in the sense of definition of this process given by Charles Galton Darwin. And the first historical undertaking, within this very logic is one by Louis de Broglie himself, just about the same time with Erwin Madelung's approach to physical interpretation of the wave function (de Broglie, 1927).

Chapter 3. De Broglie's Interpretation of Wave Function

In order to properly frame de Broglie's contribution in the order of our things here, we turn again to the same exquisite summary of Darwin already used a few times for straightening out our expression here, from which we extract yet another fragment:

> ... Now the wave aspect of matter is a century behind that of light, and so the class of experiments that have hitherto been done with electrons have not called into play any such complicated interference phenomena. For this reason *the ray problem — depending on the limitation of beams — is for electrons quite as important as the wave problem.* But that is not all, for electrons have the complication that *the wave velocity depends very strongly on the wavelength,* so that *group velocity is a very important consideration,* and the *actual motion of rays cannot be directly seen even qualitatively from the solution of the wave problem without a proper consideration of the limitation of the beam.* It thus proves more convenient not to attempt to separate the problem into two parts, *but to construct solutions of the wave equation which contain the limitations ab initio.* With this method all that remains of the ray problem is merely the derivation of the intensity from the amplitude by squaring its modulus. [(Darwin, 1927), *our Italics*]

It is quite clear that Darwin is trying to avoid the problems imposed by optical considerations, and he does that indeed in a remarkable way. It is however symptomatic, that the work from which we extracted these quotations, substantially exceeds the task suggested by its title, and we think we can pinpoint the reason. Darwin had in mind only the wave function, but the theory needs also the potential, and there is no possibility of theoretically treating it within the framework of Schrödinger equations, other than case-by-case. Which is exactly what Darwin does indeed. It is not by chance that he considers as the prototype of that "proper consideration of the limitation of the beam" the Schrödinger's work that initiated the modern idea of coherent states: as we have shown above, this aspect is required indeed for a statistical interpretation of the wave function. However, this approach raises an important issue: the wave function *is not physically interpreted, according to the very definition of interpretation* as given by Darwin himself, for the 'limitations' of the Schrödinger type do not properly cope with the idea of a light beam, and so much the less they cope with the idea of a light ray in the classical sense!

First, let us state again the grounds of what we called 'logic' at the end of the previous section. It is the concept of a Madelung fluid: *a fluid of free particles according to Schrödinger ideas*, epitomized by the time-dependent Schrödinger equation. Taken *per se*, this fluid is a continuous structure in the mathematical sense. For this kind of interpretation of the wave function, the physics needs to give it a structure, a physical structure if possible. Darwin tries to avoid the common way of appealing to optics, and clings on the pure probabilistic interpretation, which is not unnatural, but raises some strange statistical issues. Part of the problem is that the optics seems to be unable to open itself to the physical description by an ensemble, and this may justify the attitude of Darwin up to a point. However, the first known ensemble satisfying the logic just sketched in the excerpt above, was actually a physical structure related to "ray motion", "limitation of beams", "wave and group velocity" and the like, just about the time when Darwin made his point. This is the ensemble which will be taken here to indicate the necessary conditions for a general construction of such a structure within the limits of the wave mechanics, and its construction is due not to Erwin Schrödinger, but to Louis de Broglie.

Notice that the second one of the conditions (2.34) is the usual continuity equation, again, provided *the square of the amplitude of the general signal (2.28) is interpreted as a density*. That kind of density was the object of some works of de Broglie, that we have in mind for this illustration of the process of interpretation (de Broglie, 1926b,c). They were addressed to a special *optical signal*, which is not a solution of the Schrödinger equation (2.33), but of the D'Alembert equation, usually taken as the basis of physical optics. However, it turns out that de Broglie follows closely the idea of interpretation in the exact sense of the Darwin definition above, forasmuch as one can say that in optics we have to deal with *the density of a fluid conceived as an ensemble of photons*, more specifically with a modern image of the classical light ray, built on the ideas of Hooke and Newton (Mazilu & Porumbreanu, 2018).

At the time when he issued the two works just cited, Louis de Broglie was engaged in proving explicitly that there is no gap between optics and quantum theory. In its broad lines, one can say that this idea is simply a reflection of the continuity of human knowledge, which is a trait of knowledge that will often return in the course of development of the present work. The specific problem at that time was, in de Broglie's idea, to prove that the light can be seen as a flux of photons, and he intended to show that this image contradicts neither the optical nor the mechanical rules of thinking. The optical rules were considered all concentrated in the description of propagation of light, as described for the case of vacuum by the D'Alembert equation:

$$\Delta u = \frac{1}{c^2}\frac{\partial^2 u}{\partial t^2}. \qquad (3.1)$$

In this context, Louis de Broglie took note of the fact that an optical solution of the equation should be written in the form:

$$u(\mathbf{x}, t) = a(\mathbf{x})e^{i\omega[t-\Phi(\mathbf{x})]} \qquad (3.2)$$

which must then be submitted to some space constraints, evidently mandatory in optics by the presence of screens, diopters or some other obstacles met by light in space. This is exactly the philosophy outlined by Darwin, simply applied for the case of optics. As if in agreement with that philosophy, de Broglie was forced to consider the light as a fluid of particles, for the incarnation of which the best candidate was the idea of photon, floating 'in the air' so to speak, for at that moment of time the photon was just getting baptized (Lewis, 1926). We preserve here the notations of de Broglie himself, by not exhibiting the wave function in its customary notation. The main reason is that, while the signal (3.2) is clearly of the general form (2.28), it is not a solution of the Schrödinger equation (2.33). The same goes for the signal given in the equation (3.3) that follows, which is taken as *properly describing a fluid of particles*. And that is certainly due to the algebraic form of the phase, which is in a way identified with the time, a property upon which we need to return at some moment of our present work.

Indeed, by taking the light quanta as those material particles able to explain, from a classical point of view, the particulate structure of light, de Broglie noticed that one needs to assume a solution of equation (2.1), having nonetheless not only the phase, but also the amplitude time dependent:

$$u(\mathbf{x}, t) = f(\mathbf{x}, t)e^{i\omega[t-\Phi(\mathbf{x})]}. \qquad (3.3)$$

Here Φ is the same function as in (3.2), embodying the earlier idea of de Broglie that the corpuscles and the representative waves *have the same phase* (de Broglie, 1923). Why should the amplitude be variable with time here?

Louis de Broglie gives an explanation in the English version of the work (de Broglie, 1926c), and this can be summarized as follows: such an elementary particle must be described by a field satisfying the Klein-Gordon equation. This defines, choosing words of de Broglie, *the wave phenomenon called 'material point'*, and is:

$$\Delta u(\mathbf{x}, t) - \frac{1}{c^2}\frac{\partial^2 u(\mathbf{x}, t)}{\partial^2 t} = \frac{\omega^2}{c^2}u(\mathbf{x}, t); \quad \hbar\omega = m_0 c^2. \qquad (3.4)$$

The last identity here represents de Broglie's initial idea from 1923, *apparently* prompted by the relativistic mechanics, according to which one can associate, via energy, a frequency to a classical material point: the *de Broglie's frequency*. Now, the fundamental solution of this equation, based on which one can build the general solution as a linear combination according to mathematical rules, is taken by de Broglie in the general form:

$$u(\mathbf{x}, t) = \frac{a_0}{\sqrt{x^2 + y^2 + \gamma^2(z - vt)^2}} \sin \omega[t - (\beta/c)z] \qquad (3.5)$$

with a_0 a constant, $\beta \equiv v/c$ and $\gamma^2(1 - \beta^2) \equiv 1$, and the direction of motion chosen as the axis 'z' of the reference frame. No doubt, the general solution of (3.4) can be taken as being a linear combination of waves of the form (3.5). However, it should have a spacetime singularity: at the event that locates the 'material point' in motion with respect to origin of space coordinates and time, its amplitude becomes infinite. Thus, when considering the classical material point a 'wave phenomenon', if this wave phenomenon is classically located as an event, i.e. *interpreted* as in the phrasing of Darwin, the representative wave has a specific singularity at its location: its amplitude becomes infinite. In other words, by interpretation, the very concept of wave acquires here a differentia, for the particle itself gets new properties above and beyond its usual classical depiction as a position endowed with mass or charge: it a *singularity of the wave amplitude, whereby this one becomes infinite.*

However, from a 'phenomenological' point of view, we might say, the things are to be presented in a different manner. The wave is here a light wave, and it should be the space locus, in a proper geometrical sense, of the events representing the *wave phenomena called 'material points'*. The linearity of the Klein-Gordon equation allows a superposition of wave phenomena represented by (3.5) *with different velocities*, but there is a problem: as all of the material points move with the speed of light, one has $\beta \to 1$, and thus $\gamma \to \infty$, for all the waves of this type. So the resultant wave, if represented by a linear combination of such 'wave phenomena', must have rather a *vanishing amplitude* no matter where the material points representing the light are located in space and time. Thus, while classically the trajectory of the material point is, according to its definition adopted also by Darwin in the above excerpt, the locus of successive positions of a material point in motion, in a wave representation of 'the phenomenon called material point' it is simply the locus of the events where the *amplitude of the wave vanishes, no matter of the time sequence and space locations of these events*. Therefore, along the space line representing (continuously or not) a trajectory of the 'wave phenomena representing a classical material point', the phase should also be arbitrary according to this optical representation.

47

A little digression may be in order here: the phase and the amplitude of the optical elongation (3.5), which allowed the preceding speculations, are quite particular. However, only as such particulars they allowed the very construction of the special relativity, based upon D'Alembert equation (Lorentz, 1904, 1916). Thus, against these speculations one might raise the objection that the representation (3.5) is just as... special as the relativity is, and a general definition of the optical signal, as in the equation (2.28) for instance, may render them obsolete, to say the least. Two things have to be considered, however, when engaging along this line of thought. First, a signal like (2.28), satisfying any desires of generality for both amplitude and phase, is a solution of the Schrödinger equation (2.33), which proved so fruitful for the physics of the last times. If one proves that the Schrödinger equation for the free particle is vital for the natural philosophy in general, then one has to argue just 'how special' is the special relativity, viz. to give some reasons for a general... special relativity, so to speak, ideally even to find its formulation. Meanwhile notice that a stationary Schrödinger equation (2.31), deriving from (2.34), is satisfied even for the amplitude going to zero, for the phase of the signal (2.28) still can be taken as the classical action, provided the potential defined by equation (2.35) goes to zero in the position of the classical material point thus described as a 'wave phenomenon'. Secondly, as Dirac has noticed in his works which inspired a certain wave-mechanical approach to the idea of magnetic charge (Dirac, 1931, 1948), a general spatial geometric locus of zero amplitude of the signal might be instrumental for the condition of quantization based on the concept of wave function satisfying a Schrödinger equation for the free particle. Therefore, we should not avoid this line of thought, by any means. As we shall see, it is actually materialized in a certain kind of approach of the wave mechanics, leading to the *general idea of a structure of the matter.*

However, Louis de Broglie had another observation, in concordance with his own idea of phase waves. Namely, he took notice of the fact that if the amplitude function 'f'' is to have any mobile singularities, these have to move *across the surface of constant phase*, particularly normally to this surface. In this case the speed of a material point in position M at the time 't' is necessarily

$$U(\mathbf{x}, t) = - \left(\frac{\partial_t f(\mathbf{x}, t)}{\partial_n f(\mathbf{x}, t)} \right)_{M,t} \tag{3.6}$$

with the variable 'n' taken *along the very trajectory of the material point* − the symbol 'n' is here intended to suggest the idea of 'normal' to the wave surface − and the partial derivatives upon time $(\partial_t f)$ and along the normal direction $(\partial_n f)$ taken in position M at the moment 't'. Substituting (3.2) and (3.3) in equation (3.1), and making the imaginary parts of the relations thus obtained vanish, one can find the following equations connecting the optical amplitude

'A' and particle amplitude 'f' to phase Φ:

$$\frac{2}{A}\frac{dA}{dn} \equiv \frac{1}{A^2}\frac{d(A^2)}{dn} = -\frac{\Delta\Phi}{\partial_n\Phi} \tag{3.7}$$

and

$$c^2[2(\partial_n\Phi)(\partial_n f) + f\Delta\Phi] = -2(\partial_t f). \tag{3.8}$$

Then we simply have, as de Broglie notices, that the equation (3.7) will describe the diffraction phenomena *according to physical optics*, while the equation (3.8) will describe the diffraction phenomena *according to quantum theory*, i.e. *by an ensemble of particles*. It should be indeed all about diffraction, forasmuch as we have to deal here with a space locus of events distributed in space, and not with a classical trajectory *per se*. Therefore, this is indeed an *interpretation of the wave* in the acceptance of the definition given by Charles G. Darwin.

However, in the French version of his work, de Broglie assumes that if, as *one approaches at constant time* a light particle following its trajectory, the function 'f' varies as the reciprocal distance to that particle, then in the position M of the particle the *ratio between 'f' and $(\partial_n f)$ vanishes*. This fact obviously generalizes the one represented by the equation (3.5), so that it can be taken as typical for the wave mechanics. In the English version of the work, de Broglie calls for an analogy with "the spherical free point" in order to get that relation. Therefore here we would have a situation which can be characterized in general terms as a *constant time* solution of the Laplace equation in spherical coordinates. Be it as it may, the insistence of de Broglie upon this point is manifest. It can be explained by the fact that a solution is necessary *for the contradiction between the space-time singularity* where the wave amplitude jumps to infinity and *the singularity due to the classical material point image*, where the amplitude of the wave should be zero, no matter of the space-time details. Anticipating a little our explanations here, we can say that de Broglie 'felt the urge', so to speak, of completing a missing link between two scales of amplitude − infinite and infinitesimal − with a model representing the finite amplitude. The way he did this gives also an idea about the general approach to solution of such a problem. For, there is another side of this story, consistent with the ideas of wave group and phase waves, which also has a modern exquisite theoretical realization along the idea of interpretation of a wave.

The classical singularity depends exclusively on the speed of the classical material point in motion, therefore on the material point as an inertial reference frame. Now, if the light particles are indeed classical material points, for them the amplitude of the signal (3.5) vanishes, so that the phase makes indeed no sense for this signal. Obviously, however, that solution was inspired by the fact that at zero speed the equation (3.5) also represents a solution of the Laplace equation, except for the singularity in the origin. And, as we just have noticed above, starting from this equation — more precisely from Poisson equation — the Lorentz transformations were introduced (Lorentz, 1904). From this, one could also infer that the Lorentz transformations have to represent an inertial reference frame, i.e. they characterize a continuous connection between null speed and an arbitrary speed, which is indeed the case [(Boltyanskii, 1974, 1979); (Fowles, 1977)]. On the other hand, this particular requirement of Louis de Broglie is in fact universal from the point of view of the 'wave phenomenon called material point'. As we shall see here, it proves to be a necessary expression of the equivalence principle, thus making the quantum theory a fundamental component of the general relativity according to Einstein's ideas.

Fact is that under the condition of space behavior of the amplitude as requested by de Broglie, the equation (3.8) gives a special expression for the light particle velocity in a certain position; and this expression befits the classical character of phase. Indeed, using equation (3.6) and de Broglie's condition of 'approaching the point at constant time' in the form: $f/(\partial_n f) \to 0$, the formula for this velocity reveals that *the phase should be a potential of velocities*, i.e. it should have the very *classical role of the variable of action*:

$$U(\mathbf{x}, t) = c^2 (\partial_n \Phi)_{M,t}. \tag{3.9}$$

Thus, the only thing left for explanation would be the construction of a *physical light ray*, and this can be classically understood as a thin pencil of trajectories of classical material points. So, de Broglie comes to the idea that an *infinitely thin tube confining an ensemble of trajectories of light particles* would be able to do the job. Thus, the classical Newtonian image — or to be more precise, the Hookean image [see (Mazilu & Porumbreanu, 2018) for an account of historical order of development of this idea] — of the physical light ray takes, with de Broglie's description, a geometrically precise modern shape: *a generalized cylinder, whose area of any transversal section is variable with the position along the ray*.

And so it comes that de Broglie *assimilates a physical ray with a capillary tube of variable cross-section* σ, and he describes this tube by the known physical principles of the theory of capillarity. Assuming, for instance, that the flux of light particles is conserved along the ray — an assumption that can, in general, be taken as the fundamental attribute of the concept of ray within the theory of fluids — the equation representing this situation:

$$\rho U \sigma = const. \tag{3.10}$$

should be satisfied, where ρ means the Newtonian volume density of the particles of light. Taking the logarithmic derivative in the direction of the ray, one can find

$$\frac{1}{\rho}\frac{\partial \rho}{\partial n} + \frac{1}{U}\frac{\partial U}{\partial n} + \frac{1}{\sigma}\frac{\partial \sigma}{\partial n} = 0. \tag{3.11}$$

By a "known theorem of geometry", as de Broglie declares, one can calculate the last term here. Now, because the physical ray is a space construction, it would be hard to decide the meaning of $\partial/\partial n$ − is it effectively variation along the ray itself, or along the normal to the wave surface as de Broglie assumes!? − but to a good approximation we can take that it means variation along the normal, to start with. It is, indeed, only in this case that we can take advantage of that 'known theorem' to which de Broglie alluded, and according to which the last term in (3.11) is the double of the mean curvature of the surface $\Phi = const.$ in a given position. That quantity has as expression the sum of the principal curvatures of the surface:

$$\frac{1}{R_1} + \frac{1}{R_2} = \frac{\Delta\Phi - \partial_n^2\Phi}{\partial_n\Phi}. \tag{3.12}$$

Here $R_{1,2}$ are the radii of curvature of the principal sections of the surface. With (3.9) and (3.12), the equation (3.11) now takes the form

$$\frac{1}{\rho}\frac{\partial\rho}{\partial n} = -\frac{\Delta\Phi}{\partial_n\Phi} \tag{3.13}$$

and comparing this with (3.7), one finds

$$\rho = const \cdot A^2. \tag{3.14}$$

Thus, Louis de Broglie has the essential result of interpreting the physical optics based on diffraction phenomena *without making any reference to the idea of harmonic oscillator* and its classical dynamics in order to calculate the intensity of light. Indeed, the equation (3.14) shows that the density of the light particles classically conceived as material points or localized quanta, is proportional with the intensity of the classical theory of light. The interference phenomena are, therefore, explained by the corpuscular theory, just as well as the diffraction phenomena, provided we add to the wave mathematical image a necessary property deriving from the wave representation of classical material point: the ratio between the amplitude of the wave and the normal derivative of its wave surface, taken at constant time, vanishes in the position of the material point.

An issue still lurks in the background though, and a fundamental gnoseological issue for that matter: was this effort of mathematics and stretch of imagination necessary at all for our knowledge? From the point of view of the continuity of the knowledge, the answer is definitely affirmative. Indeed, making reference to the harmonic oscillator in the case of light − in order to interpret the intensity of light, for instance, to say nothing of some other physically fundamental necessities − is, stretching a little the meaning of word, 'illegal'. For, as a purely dynamical system, the harmonic oscillator is a dynamical system described by forces proportional with displacements (Hooke-type elastic forces), and in the case of physical optics the second principle of dynamics is only incidental, being introduced only by a property of transcendence of the second order ordinary differential equation: it describes any type of periodic processes. And the fact is that, in the foundations of modern physical optics, the periodic processes of diffraction have more to do with the theory of statistics than with the dynamics (Fresnel, 1827).

This is, however, not to say that the harmonic oscillator is to be abandoned altogether, as a model, because this is not the case, either from experimental point of view or theoretically. All we want to say is that we need to find its right place and form of expression in the theory, and this is indicated again through the order imposed by the measure of things, this time as their mass. Indeed, dynamically, the second order differential equation involves a finite mass. On the other hand, for light the mass is evanescent, and if the second order differential equation is imposed by adding the diffraction to the phenomenology of light, this means that it describes actually *a transcendence between finite and infinitesimal scales of mass.* As we shall see later, the mathematics of scale transitions between finite and infinitesimal in SRT, respects entirely rules related to the harmonic oscillator model. In fact, the whole wave mechanics as a science can be constructed based on such rules, which appear to be universal.

The Appropriate Geometry of de Broglie's Idea

The effectively inadequate point of the de Broglie's argument is in fact only the mathematical expression of his way to conclusions, for, as we have seen, everything else is in the right place. So, we assume now the burden of a first mathematically appropriate expression of de Broglie's results, in order to see where they lead us. The first step suggested in his argument would then be construction of an adequate geometrical theory of the implicitly defined surfaces in space. [(Hughes, 2003); (Goldman, 2005)]. Louis de Broglie works only with this kind of surfaces, but from a particular point of view. We shall give here only an inventory of the geometrical formulas necessary in reaching the equation (3.12), because this equation is the fundamental result which represents the key of argument, and thus motivates its conclusions. This result is simply borrowed by de Broglie from the classical theory of capillarity (Poincaré, 1895). There, indeed, it is only presented in the particular one-dimensional form used by de Broglie, but it can be presented in a general three-dimensional form.

The equation of a family of parallel surfaces like, for the case in point, the waves in free space for instance, can be written parametrically in the form:

$$\mathbf{r}(u,v) = \mathbf{x}(u,v) + \lambda \cdot \hat{\mathbf{e}}_3(u,v). \qquad (3.15)$$

Here (u,v) are the parameters on the surface chosen as basis, $\hat{\mathbf{e}}_3$ is the common normal of the family, and $\mathbf{x}(u, v)$ represents the basis surface. In order to be in agreement with our current subject, the parameter λ might mean, for instance, the wavelength, like in the geometrical optics, so that the equation (3.15) can be taken as suggesting the fact that when we have to deal with a family of successive wave surfaces, the distance along the normal between them is the wavelength. Let us assume that this is a constant, in order to simplify the argument: so, if there is a dispersion of the vaves our assumption says that it is not accompanied by a variation of the wavelength. In this case, from (3.15) we have by differentiation:

$$d\mathbf{r}(u, v) = d\mathbf{x}(u, v) + \lambda \cdot d\hat{\mathbf{e}}_3(u, v). \qquad (3.16)$$

Using now some equations characteristic to the differential geometry of surfaces, we further have that the first fundamental form of a generic surface of the family (3.15) is given by the vector

$$|s_\lambda\rangle = \mathbf{H} \cdot |s\rangle; \quad \langle s| \equiv (s^1, s^2); \quad \mathbf{H} \equiv \mathbf{1} - \lambda\mathbf{h} \qquad (3.17)$$

where $|s\rangle$ is the elementary displacement in the surface and $\mathbf{h}$ is an *ancillary matrix* used in classical theory of surfaces to represent the variations of curvature, or even curvature properties themselves (Mazilu & Agop, 2015). In the present context it plays the part of a curvature matrix, and allows us to say that there is a local linear homogeneous relation between the different wave surfaces of this linear family. Thus the element of an oriented generic surface of the family, located by the parameter λ, is given by an exterior differential 2-form, which can be represented as the exterior product [(Stoker, 1989), p.352]

$$s_\lambda^1 \wedge s_\lambda^2 = (s^1 \wedge s^2)(1 - 2H\lambda + \lambda^2 K). \qquad (3.18)$$

Here H is the *mean curvature* of the reference surface, while K is its *Gaussian curvature*. Using this, we can calculate the mean and Gaussian curvatures of a generic surface of the family. This can be done by using two differential relations which, for the reference surface are [(Stoker, 1989), pp.352ff]

$$\begin{aligned}
(d\mathbf{x}) \times^\wedge (d\hat{\mathbf{e}}_3) &= 2H(s^1 \wedge s^2)\hat{\mathbf{e}}_3, \\
(d\hat{\mathbf{e}}_3) \times^\wedge (d\hat{\mathbf{e}}_3) &= 2K(s^1 \wedge s^2)\hat{\mathbf{e}}_3.
\end{aligned} \qquad (3.19)$$

The symbol '$\times^\wedge$' means here a vector product of the vectors having differential 1-form as components, where the monomials are defined by the exterior multiplication of 1-forms instead of regular multiplication of numbers. For a generic surface of the family (3.15) we have the equivalent of equation (3.19) as

$$\begin{aligned}
(d\mathbf{r}) \times^\wedge (d\mathbf{e}_3) &= 2H_\lambda(s^1 \wedge s^2)\hat{\mathbf{e}}_3, \\
(d\mathbf{e}_3) \times^\wedge (d\mathbf{e}_3) &= 2K_\lambda(s_\lambda^1 \wedge s_\lambda^2)\hat{\mathbf{e}}_3.
\end{aligned} \qquad (3.20)$$

because the normal direction is common to the family. Using the equations (3.16) and (3.18), the two measures of the curvature of a generic surface are given as function of the wavelength by the following relations:

$$H_\lambda = \frac{H + \lambda K}{1 - 2\lambda H + \lambda^2 K}, \quad K_\lambda = \frac{K}{1 - 2\lambda H + \lambda^2 K}. \quad (3.21)$$

The "known theorem" used by de Broglie in calculating the last term in equation (3.11) can be obtained from equation (3.21) for a finite portion of the generic surface of the family. Indeed, taking (3.18) for expressing the difference between σ_λ and σ, we have

$$\frac{1}{\sigma}\frac{d\sigma}{dn} \equiv \frac{1}{\sigma}\lim_{\lambda \to 0}\frac{\sigma_\lambda - \sigma}{\lambda} = -2H;$$

$$\sigma \equiv \iint s^1 \wedge s^2, \quad \sigma_\lambda \equiv \iint s_\lambda^1 \wedge s_\lambda^2 \quad (3.22)$$

because λ can be taken as an affine parameter along the normal of the family. For completeness, we need to establish now the relation (3.12) in its general form, but for a *surface defined implicitly*, as in the case of Louis de Broglie.

To this end we have to consider the definition of the normal to a surface defined implicitly by equation $\phi = constant$. Such a normal is defined by vanishing of the differential form:

$$d\mathbf{x} \cdot \nabla\phi \equiv dx^k \phi_k = 0 \quad \therefore \quad \hat{\mathbf{e}}_3 = \frac{\nabla\phi}{|\nabla\phi|} \quad (3.23)$$

which represents the tangent plane to surface, and expresses the natural fact that if $d\mathbf{x}$ is taken as a virtual displacement *in the surface*, the normal has to be perpendicular to it. After a longer, but otherwise direct calculation, the differential $d\mathbf{e}_3$ of the normal unit vector − the so-called *Weingarten application* − turns out to have the components

$$dn_i = \Psi_{ij}dx^j \text{ with } (\Psi_{ij}) \equiv \mathbf{\Psi} = \frac{\mathbf{H}(\phi)}{|\nabla\phi|} - \frac{\mathbf{H}(\phi) \cdot (|\nabla\phi\rangle\langle\nabla\phi|)}{|\nabla\phi|^3} \quad (3.24)$$

where $\mathbf{H}(\phi)$ is the Hessian matrix of the function ϕ. Thus, the second fundamental form of such a surface is given by

$$\hat{\mathbf{e}}_3 \cdot d^2\mathbf{x} = -d\hat{\mathbf{e}}_3 \cdot d\mathbf{x} = -\Psi_{ij}dx^i dx^j \quad (3.25)$$

which would indicate that the mean curvature can be given by the trace of the matrix $\mathbf{\Psi} \equiv (\Psi_{ij})$. This is what happens indeed [(Goldman, 2005), equation (4.2)]:

$$-2H = \text{Tr}(\mathbf{\Psi}) = \frac{\Delta\phi}{|\nabla\phi|} - \frac{\langle\nabla\phi|\mathbf{H}(\phi)|\nabla\phi\rangle}{|\nabla\phi|^3}. \quad (3.26)$$

This should be the generalization of expression from equation (3.12), which therefore enables us to establish the following correspondences:

$$\frac{\partial \phi}{\partial n} \leftrightarrow |\nabla \phi| \quad \text{and} \quad \frac{\partial^2 \phi}{\partial n^2} \leftrightarrow \frac{\langle \nabla \phi | \mathbf{H}(\phi) | \nabla \phi \rangle}{|\nabla \phi|^2}. \tag{3.27}$$

Thus the derivative along the normal in the case considered by Louis de Broglie must be the magnitude of the gradient of phase surface, while the second order derivative along the normal must be the component of the Hessian matrix of the surface along its normal. Starting here, a further generalization should be in order from physical point of view − or, better, from a fractal point of view, because in our opinion the fractal point of view becomes, by Nottale's ideas, a physical point of view − namely the description of a fragment of a surface. Obviously, this can be done by generalizing the idea of curvature and metric to satisfy the physical definition of a surface, which is one of our main points here.

Meanwhile, for completeness of the present image of de Broglie's theory, we just transcribe here the Gaussian curvature, in the framework of the implicit definition of a surface. It is given by a little more complicated formula [see (Goldman, 2005), equation (4.1)]:

$$K = \frac{\langle \nabla \phi | \mathbf{\Phi}(\phi) | \nabla \phi \rangle}{|\nabla \phi|^4} \tag{3.28}$$

where $\mathbf{\Phi}$ is the reciprocal of the Hessian matrix $\mathbf{H}$ of surface, defined by:

$$\mathbf{\Phi}(\phi) \equiv \det[\mathbf{H}(\phi)] \cdot \mathbf{H}^{-1}(\phi). \tag{3.29}$$

However, as we have shown above, this relation is not used by Louis de Broglie, who simply follows the particular classical line according to which the surface tension − like the fluid surface itself in a capillary tube − is exclusively defined by its mean curvature. This item calls, again, for a generalization of the theory to the physical definition of fragment of surface.

Lessons and Mandatory Developments

It is the time to summarize the achievements with respect to the Schrödinger theory, and then make an inventory of the problems that remain to be solved, and for which SRT offers ideas of solution. It is by now clear that only the theory of fluids can offer the image of a structure of matter, *an interpretation* we should say, in order to devote ourselves to an established language, whatever the time and space scale of its location. That explains the intervention of Erwin Madelung, who maintained the Schrödinger initial point of view, but introduced a special kind of continuity of matter, connected with the Newtonian definition of its density. Madelung's idea has received a physical elaboration only under the assumption that between Schrödinger wave function and the Newtonian-type density of matter there is a quadratic relationship: the density is the square of the modulus of the "essentially complex" wave function. Quite naturally then, the theory should be limited to places where the matter exists continuously, but this raises problems related to the physical description of such places. In turn, these problems impose choices about the elementary units of a physical structure, and the Madelung's choice is indicated by him with no hesitation: for instance, the electrons "penetrate each other without merging". In other words, the electron, by then an elementary particle — which, as a matter of fact, remained elementary ever since — *is the home of the continuous matter*, once it behaves inside other electron like the matter in ether. For, the ether, was then, as it is nowadays, the quintessential matter, which the bodies penetrate "without merging". So, if we ask how to describe it as a physical structure, we actually have no space-time possibility: there is no possibility of "merging" indeed, in order to assure a physical structure. And thus, the image of the matter should be that of a continuum having "an appearance suggesting a vivid swarm, whose particles have infinite free paths". This is the first remark ever upon the fact that the only kind of structure of the matter *is not a physical structure*, but only a mathematical one, resulting from the theory of ensembles via the continuum hypothesis.

Enters now Louis de Broglie, with the only known candidate to an ensemble of classical material points to date: *the light in a Huygens-type interpretation*. He could see the material points as photons, in view of his recent discovery that the frequency can be associated with the rest energy of a classical material point (de Broglie, 1923). However, de Broglie did not describe the light globally, as Huygens' description of the light requires, but rather locally, much in the manner of Fresnel, in order to account for the difraction phenomena (Fresnel, 1827). While this specific approach of the idea of an ensemble describing the matter seemed to leave aside the Madelung's problem of 'penetrating without merging', it showed that the quadratic correlation assumed by Madelung in reaching such a conclusion is actually only a local correlation. Thus de Broglie added an important attribute to the very classical notion of a light ray, as we duly noticed before. However, his theory means in fact much more, far and beyond the physical theory of light.

First of all, locally de Broglie had to assume that if, *as one approaches at constant time* a light particle following its trajectory, the amplitude function 'f' of the wave associated to the representative classical material point varies as the reciprocal distance to that point, then in the very position of the point the *ratio between 'f' and $(\partial_n f)$ vanishes*. It is, of course, necessary to explain what that 'constant time' means, and insofar as the idea of 'approach' here suggests a motion, we can appeal to intuition. Intuitively, therefore, a 'motion of approach' of de Broglie's type, if it ever exists, should be way faster than the motion of the material point itself along the ray, in order to appear as instantaneous at the time scale of this last motion. This is not as unusual as it seems, at least in physics, and can even be framed into a *general rule of describing the matter:* a regular motion of a physical structure should be 'adiabatic' with respect to internal motions explaining the 'wave phenomenon called material point'. For once, this is the gist of the duality between wave and particle, but it goes way beyond the theory of relativity. Which is what might be able to explain the universal success of de Broglie's theory as an incentive of approaching the wave mechanics, in the first place. Because, as a matter of fact, the theory of relativity is a strong point against de Broglie's theory of relation between phase waves and group of waves. As Charles Galton Darwin puts it:

> When de Broglie first developed his wave theory *he based it largely on the help of relativity.* The consequence is that the *wave velocity of an electron is much greater than the velocity of light.* This is ultimately correct of course, but *it is an unnecessary complication always to have to consider relativity in dealing with quite slow motions.* We shall throughout the present work avoid doing so by taking a factor
>
> $$\exp\{-i(2\pi/\hbar)mc^2 t\}$$
>
> out of our wave functions, which is done by a simple and familiar modification of the wave equation. We shall, of course, get quite a different value for the wave velocity from that of de Broglie. To borrow an analogy from the practice of wireless telephony, *we are observing our waves with the help of a heterodyne frequency mc^2/h,* and *when we speak of the phase of a wave we mean the phase of the sound heard in the telephone, not that of the aethereal vibrations.* [(Darwin, 1927); *our Italics*]

The fact that a velocity greater than the limit velocity does not make physical sense is indeed true, provided the limit velocity is that of light, as in physical optics. In this case the interpretative ensemble would have as constituents material points moving with a speed greater than that of light. But this is not always the case in physics, and the *de Broglie's physical ray* allows us to construct an interpretation based on the idea of adiabaticity mentioned above. The specific case of the classical hydrogen atom, from which, when it comes to quantum electrodynamics, it all started, enables indeed a ray interpretation in the sense of de Broglie, when this model is taken in its utmost mathematical generality.

Therefore, in order to show what is the specific key point of such a construction, let us notice that de Broglie's approach to the concept of physical ray of light offers a natural way of description of the classical nuclear atom. In fact, not quite a description of the whole atom, but only of a Madelung electron, virtually capable of 'penetrating another electron without merging it'. All it takes for realizing that this possibility is unraveled by de Broglie's idea, is to notice that the whole Keplerian orbit in the physical structure of a classical nuclear atom can be actually *assimilated to a capillary tube*. Indeed, if the electron is regarded as a classical material point described as a 'wave phenomenon', then it should be a spatially extended structure, within the space of which those de Broglie 'motions of approach' are much faster than the velocity of motion around the nucleus. With respect to those *fictitious* motions, the proper electron motion around nucleus appears as 'adiabatic' by its slowness. The whole region occupied by this structure then describes an 'elliptic ray', as it were, physically realized as an 'elliptical capillary tube' around the nucleus. In fact, with a little geometry, this image shows a lot more than it appears at the first sight.

Assume, indeed, that the region occupied by the wave phenomenon describing the electron as a classical material point is of a regular shape: an ellipsoid say, even of variable dimensions as it goes around the nucleus. Then the de Broglie's capillary tube can be described as the envelope of the set of instantaneous ellipsoidal shapes assumed by the electron in its journey around nucleus. This is, in fact, a generalized Huygens principle. The generalization refers to the fact that the principle can be applied to matter exactly the same way it is applied to light. This means that the capillary tube is actually a wave surface, comprising a space region only within which the fictitious de Broglie particles have the freedom to move.

Out of this whole summary of classical achievements, we have an inventory of mathematical developments necessary in order to initiate a physical theory:

The first comes the Kepler problem: where do we stand with the Kepler problem? Of course, we all know the actual standing, but that standing has nothing to do with the Madelungs fluid or de Broglie's *wave-particle duality*. And a proper theory of matter expressly requests these concepts: the matter cannot be defined, for the benefit of physics of course, but only through interpretation and, as we have shown above, the interpretation cannot be accomplished without them.

Secondly, these very concepts ask, in turn, for a proper further development of the theory of surfaces: a theory of *surfaces made out of physically defined fragments*. Because the physics cannot define a surface but through its perceived fragments. The quintessential example is the contribution of the phenomenology of diffraction to the idea of wave surface: the Fresnel reconstruction of the wave surface from 'infinitesimal fragments', does not lead exactly to a Huygens wave surface. However, insofar as the theory of de Broglie is mandatory in properly completing the Fresnel's theory of light into a physical theory, a point that should be duly taken by theoretical physics, it asks for *a proper generalization of the idea of capillarity*: this one should be part an parcel of a *physical theory of surfaces*. Thus, we do recognize it as one of our mathematical tasks with the present endeavor. As we shall see, it is closely connected to statistics and stochasticity of the processes in matter.

Chapter 4. The Planetary Model as a Dynamical Kepler Problem

The classical Kepler motion is the usual model describing either the planets re-volving around the Sun or the electrons revolving around a nucleus, within the framework of classical dynamics. In the spirit of Nottale's conclusions quoted in our Introduction to this work, we will insist here in detail on the fact that this problem shares with the special relativity one distinctive feature: *the limita-tion in magnitude of the initial velocities possible in a certain instance.* This is quite a general feature, fundamental for the quantitative knowledge we should say, that makes out of special relativity a... general theory, and bestows upon de Broglie's theory of wave-particle duality the necessary status of indepen-dent physical theory. In hindsight, this is nothing short of SRT! For the mo-ment, however, we limit our considerations to geometry, pointing out that the mathematics of the classical Kepler problem allows us to construct a specific non-Euclidean geometry, which can be shown to arise quite naturally from the requirement that the Kepler orbit should be closed [(Belbruno, 1977); (Milnor, 1983)]. In what follows we keep the description as close as possible to the intu-itive and classical aspect of the problem, in order to best unveil what we believe is its true physical nature.

The classical space image of a Kepler motion can be dynamically explained via Newtonian equations of motion, with a central force having a magnitude inversely proportional to the square of the distance between the center of at-traction and the attracted classical material point. In vector notation these equations are:

$$\ddot{\mathbf{r}} + \frac{\kappa}{r^2}\frac{\mathbf{r}}{r} = \mathbf{0}. \tag{4.1}$$

Here κ is a physical constant, $\mathbf{r}$ denotes the position vector of the classical material point whose motion is calculated with respect to the center of force, and an overdot means derivative with respect to time, as usual. The constant κ may or may not depend on quantities related to the material point in motion, depending on the Newtonian forces involved in its dynamics. For instance, in the case of motion in pure gravitational field, this constant does not contain but the properties of the material point considered the source of the gravitational field (the gravitational mass). It is only in electric and magnetic problems, as in the case of planetary atom, that κ contains also properties of the revolving material point thus described by our Kepler problem. This situation is simply a reflection of the fact that, in the first place, the second principle of dynamics confers a special position to the inertial mass and, secondly, the inertial mass itself is identified with the gravitational mass.

We can simplify the algebra leading to solution of equation (4.1) by restricting the geometry to the plane of motion, where the position vector will be denoted by $\mathbf{x}$, which is in fact all we need for the argument of the present work. We can do this as a benefit of the fact that we are dealing with a central force, and thus the motion is plane. If the generic coordinates of the point in motion with respect to the center of force are ξ and η say, the equation (4.1) is then equivalent to the system (Mittag & Stephen, 1992):

$$\ddot{\xi} + \frac{\kappa}{x^2}\cos\phi = 0; \quad \ddot{\eta} + \frac{\kappa}{x^2}\sin\phi = 0. \tag{4.2}$$

Here $x \equiv |\mathbf{x}|$ and ϕ, are the polar coordinates of the moving point in the plane of motion with respect to the attraction center. The magnitude of rate of area swept by the position vector is a constant of motion given by

$$\dot{a} \equiv \xi\dot{\eta} - \eta\dot{\xi} = x^2\dot{\phi}. \tag{4.3}$$

This area rate, a constant of motion according to the second of the Kepler laws, allows us an elegant integration of the system (4.2) with the analytical form of the trajectory as a direct outcome. First we define the complex variable

$$z \equiv \xi + i\eta = re^{i\phi}$$

so that (4.2) can be written in the form

$$\ddot{z} + \frac{\kappa}{x^2}e^{i\phi} = 0.$$

Now, use (4.3) to eliminate x^2, such that

$$\ddot{z} + \frac{\kappa}{\dot{a}}e^{i\phi}\dot{\phi} = 0 \quad \therefore \quad \dot{z} = i\left(\frac{\kappa}{\dot{a}}e^{i\phi} + v\right), \tag{4.4}$$

where $v \equiv v_1 + iv_2$ is a complex constant of integration to be defined by some initial conditions of the problem. Then the analytical equation of motion can be extracted directly, by calculating the area constant (4.3) with the help of the first result of integration given in equation (4.4). In polar coordinates the final result is:

$$\frac{\dot{a}}{x} = \frac{\kappa}{\dot{a}} + v_1\cos\phi + v_2\sin\phi. \tag{4.5}$$

The shape of this trajectory is nevertheless best pictured, in well-known details, by switching back to the Cartesian coordinates ξ and η, thus obtaining instead of (4.5) the second-degree plane curve − a *conic*:

$$\left(\frac{\kappa^2}{\dot{a}^2} - v_1^2\right)\xi^2 - 2v_1 v_2 \xi\eta + \left(\frac{\kappa^2}{\dot{a}^2} - v_2^2\right)\eta^2 + 2\dot{a}(v_1\xi + v_2\eta) = \dot{a}^2. \quad (4.6)$$

The first known detail is that the center of this conic *is not the center of the force*, i.e. in our chosen coordinates it is not located in the origin of the plane of motion. However, we can make this statement more precise: with respect to the center of force, the center of orbit has the coordinates:

$$\xi_0 = -\frac{\dot{a}v_1}{\Delta}, \quad \eta_0 = -\frac{\dot{a}v_2}{\Delta}; \quad \Delta \equiv \frac{\kappa^2}{\dot{a}^2} - v_1^2 - v_2^2. \quad (4.7)$$

Some other well-known details can further be made precise as follows. Assuming the center of the conic at finite distance with respect to the center of force, and referring the trajectory to this center by a translation of vector $\xi_0, \eta_0 : x = \xi - \xi_0, y == \eta - \eta_0$, its equation becomes

$$\left(\frac{\kappa^2}{\dot{a}^2} - v_1^2\right)x^2 - 2v_1 v_2 xy + \left(\frac{\kappa^2}{\dot{a}^2} - v_2^2\right)y^2 = \frac{\kappa^2}{\Delta}.$$

The quadratic form from the left hand side of this equation is completely described from a geometrical point of view, by the 2×2 special matrix

$$\mathbf{a} \equiv \frac{\Delta}{\kappa^2}\begin{pmatrix} \frac{\kappa^2}{\dot{a}^2} - v_1^2 & -v_1 v_2 \\ -v_1 v_2 & \frac{\kappa^2}{\dot{a}^2} - v_2^2 \end{pmatrix}. \quad (4.8)$$

The metric elements of the trajectory, i.e. its semiaxes, are given by eigenvalues of inverse of this matrix, which can be written in the form:

$$\mathbf{a}^{-1} = \frac{\dot{a}^4}{\kappa^2}\frac{1}{(1 - e_1^2 - e_2^2)^2}\begin{pmatrix} 1 - e_2^2 & e_1 e_2 \\ e_1 e_2 & 1 - e_1^2 \end{pmatrix} \equiv \frac{\dot{a}^4}{\kappa^2}\frac{1}{\lambda^4}(\lambda^2 \mathbf{1} + \mathbf{e} \otimes \mathbf{e}). \quad (4.9)$$

Here $\mathbf{1}$ is the 2×2 identity matrix, while the vector $\mathbf{e}$ and the scalar λ are defined by:

$$e_1 = \frac{\dot{a}}{\kappa}v_1; \quad e_2 = \frac{\dot{a}}{\kappa}v_2; \quad \lambda \equiv \sqrt{1 - e_1^2 - e_2^2}. \quad (4.10)$$

Further on, the eigenvectors of this matrix are $\mathbf{e}$ and a vector orthogonal to $\mathbf{e}$ in the plane of motion, so that the corresponding eigenvalues are given by:

$$a^2 = \frac{\dot{a}^4}{\kappa^2\lambda^4}, \quad b^2 = \frac{\dot{a}^4}{\kappa^2\lambda^2}. \quad (4.11)$$

Thus, the *eccentricity* of the trajectory, defined geometrically by

$$e^2 \equiv \frac{a^2 - b^2}{a^2} = 1 - \lambda^2 \quad \therefore \quad e^2 = e_1^2 + e_2^2$$

turns out to be the length of vector $\mathbf{e}$, which explains in fact our very notation. Again, by equation (4.7) the relative coordinates of the center of trajectory with respect to its focus, or vice versa, can be written as:

$$\xi_0 = -\frac{\dot{a}^2}{\kappa} \frac{e_1}{1 - e_1^2 - e_2^2}, \quad \eta_0 = -\frac{\dot{a}^2}{\kappa} \frac{e_2}{1 - e_1^2 - e_2^2}. \tag{4.12}$$

Now, the ratio of these two coordinates gives the *orientation of the orbit*, so that if θ is the angle of this orientation we may write the components of the eccentricity vector in the form

$$e_1 = e \cos \theta, \qquad e_2 = e \sin \theta. \tag{4.13}$$

These are all well-known facts, as we said, only expressed in a mathematical form that makes the role of the necessary initial conditions more obvious for our purpose: on one hand they control the shape of the orbit while, on the other hand, they delimit the extension of the area covered by the very center of force, which is measured by the eccentricity vector.

It is now the time to start with less known things about the geometries involved in the Kepler motion. The first such thing to be noticed is the tensor involved in the relation (4.9), viz.

$$\mathbf{g} \equiv \frac{1}{\lambda^4}(\lambda^2 \mathbf{1} + \mathbf{e} \otimes \mathbf{e}). \tag{4.14}$$

This matrix is the subject of a surface embeding theorem [see (Hu & Zhao, 1997), Proposition 1]: the matrix $\mathbf{g}$ is the *first fundamental form* of a hyperbolic twofold having curvature -1, immersed into a hyperbolic threefold of curvature 'c'. The *second fundamental form* of the twofold is given by the 2×2 matrix having the entries

$$h_{ij} \equiv \frac{\partial^2 u(\mathbf{e})}{\partial e_i \partial e_j}.$$

Here u(e) is a smooth solution of the Monge-Ampère equation:

$$\det \mathbf{h} = -\frac{1+c}{(1-e^2)^2} \tag{4.15}$$

which realizes the embedding. In other words, we are entitled to consider the region around the center of force in the classical Kepler problem as a *hyperbolic*

63

space of constant curvature. In view of the definition (4.10) for the coordinates, this is basically *a velocity space*: its coordinates are defined by the velocity chosen as initial condition of the Kepler orbit. On the other hand, the presence of the Hessian matrix as the second fundamental form, is indicative of the pertinence of an implicit theory of surfaces as in the case of theory of Louis de Broglie. Before any further elaboration along these lines, let us uncover a few more things connected with this very point.

We can rewrite the metric (4.14) in a well-known form, by recalling that for elliptic trajectories 'e' is confined to values taken in the interval between -1 and +1, so that the change of parameter

$$e = \tanh \psi \qquad (4.16)$$

with ψ real, is legitimate. With this, the metric defined by (4.14) becomes

$$(ds)^2 = (d\psi)^2 + \sin^2 \psi (d\theta)^2. \qquad (4.17)$$

from which it is obvious that we have to deal with a metric of negative curvature, as the embedding theorem states. Again, we have to deal here with a property of the velocity space in special relativity, forasmuch as the quadratic form (4.17) is simply the metric of the space of relativistic velocities (Fock, 1959), provided we identify d with the arclength of the regular unit sphere. This can be taken, and we shall take it here indeed, as the basis of the possibility to 'render the theory relativistic' as Nottale puts it.

For now, we are interested here only in a suggestion correlated to this idea, coming out from the previous considerations as follows. From equation (4.7) and (4.11) we can get the position of the center of force with respect to the center of orbit which, in (ψ, θ) coordinates. is given by:

$$-\xi_0 = \frac{\dot{a}^2}{2K} \sinh 2\psi \cos \theta; \qquad -\eta_0 = \frac{\dot{a}^2}{2K} \sinh 2\psi \sin \theta. \qquad (4.18)$$

This means that if the orbit is fixed, say by a kind of gauging, the variables (ψ, θ) describe the *a priori* variation of the position of the center of force compatible with that orbit. In other words, ξ_0 and η_0 can be considered as possible *coordinates of the center of force in the plane of orbit*, when this one has a fixed center. In this capacity, as we have already noticed above, they are proper coordinates in the space region occupied by the center of force, i.e. the interior of the atomic nucleus, Sun, galactic nucleus, and the like. That space region is then described by a Lobachevsky-type geometry, which can be presented as a Cayleyan geometry with respect to a sphere, taken as an *absolute of space* containing the matter in Kepler problem.

With this observation we come to the main question: how is this region to be described *in space*? More to the point, how is the *past velocity field* of

64

initial conditions correlated with a *contemporary space position*? The answer is immediately at hand through a suggestion given by a known mathematical construction. Indeed, the eccentricity vector from equation (4.13) has a direct connection with the theory of classical potentials *via harmonic maps*. This raises an *important issue* which may affect the way we conceive the necessary stochasticity of matter. Indeed, in order to exhibit the connection just mentioned let us notice that we may have a few possibilities of constructing harmonic maps between the usual space and the twofold described by the metric (4.17). One of these is to consider the complex variable, 'z' say, written in terms of parameters (e, θ) in the form for which that metric is given by a well-known formula:

$$z \equiv i \frac{\cosh(\psi/2) + \sinh(\psi/2)e^{-i\theta}}{\cosh(\psi/2) - \sinh(\psi/2)e^{-i\theta}}; \quad (ds)^2 = -4\frac{dzdz^*}{(z - z^*)^2}. \quad (4.19)$$

The metric is obviously a metric of the Lobachevsky plane in the Poincaré disk representation. As it happens, this complex parameter even represents a harmonic map from the usual space into the Lobachevsky plane having the metric (4.19), *provided ψ is a solution of the Laplace equation in free space*. Perhaps we need to sketch the proof in order to see how this statement is understood.

Indeed, the problem of harmonic correspondences between space and the hyperbolic plane is described by the stationary values of the energy functional corresponding to the metric (4.17). This is defined as the volume integral of an integrand obtained from that metric by transforming the differentials into space gradients [(Eells & Sampson, 1964); (Misner, 1978)]. The stationary values of energy functional, therefore, correspond to solutions of the Euler-Lagrange equations for a metric Lagrangian like

$$-4\frac{(\nabla z) \cdot (\nabla z^*)}{(z - z^*)^2}, \quad (4.20)$$

where the gradient is purely Euclidean. These are given by

$$(z - z^*)(\nabla^2 z) - 2(\nabla z)^2 = 0 \quad (4.21)$$

and its complex conjugate. Then it is easy to see, by a direct calculation, that 'z' from equation (4.19) verifies equation (4.20) when ψ is a solution of Laplace equation, *and θ is arbitrary*, in the sense that *it does not depend on the position in space*. Nevertheless, it might depend on the local time of the original Newtonian dynamics; but as long as the parameter ψ is real, θ should not depend on space coordinates.

The issue mentioned above can now be made obvious by the fact that, while in equation (4.19) θ is just *incidental*, one cannot say the same if we apply the principle of harmonic mapping to the metric (4.17) directly. Indeed, the harmonic map problem for the metric (4.17) is provided by the solution of the couple of partial differential equations

$$\nabla^2 \psi - \sinh \psi \cosh \psi (\nabla \theta)^2 = 0; \quad \nabla[\sinh^2 \psi (\nabla \theta)] = 0. \qquad (4.22)$$

This time the *orientation of the orbit is not a matter of arbitrary choice* with respect to the position in space, because it depends on space coordinates. However, the arbitrariness still exists, but this time it is moved to another level, so to speak, namely that of the constants of integration of the equations (4.22). In order to show a possible connotation in that direction, let us briefly present a solution of these equations.

From the form of equations (4.22) it becomes clear that we should have to deal with some solitonic-type solutions in space. As, incidentally, the solution should be plane, it can be obtained by taking the natural assumption that ψ and θ depend on the position in space via the distance from a certain plane, therefore through a linear form:

$$\xi \equiv \mathbf{a} \cdot \mathbf{x} \qquad (4.23)$$

where $\mathbf{a}$ is an arbitrary constant vector. Then, assuming further that this vector is non-null, the equations (4.22) take the form

$$\psi'' - \sinh \psi \cosh \psi (\theta')^2 = 0; \quad [\sinh^2 \psi (\theta')]' = 0 \qquad (4.24)$$

with a prime denoting derivative with respect to ξ. The second of these equations can be integrated right away, and gives the result

$$\theta' = \frac{C}{\sinh^2 \psi} \qquad (4.25)$$

where C is a constant of integration. Using (4.25) in the first of equations in (4.24) gives

$$\psi'' = C^2 \frac{\cosh \psi}{\sinh^3 \psi}$$

which can again be integrated directly, multiplying it by ψ'. This leads to the first implicit integral

$$\psi'^2 + \frac{C^2}{\sinh^2 \psi} = C_1^2.$$

All things considered real, ψ' can be written in the form

$$\psi' = \pm \frac{\sqrt{C_1^2 \sinh^2 \psi - C^2}}{\sinh \psi}$$

66

and this equation can be rearranged in a form with total differentials

$$\frac{sinh\psi d\psi}{\sqrt{\sinh^2\psi - \sinh^2\psi_0}} = \pm C_1 d\xi; \quad C_1^2 \sinh^2\psi_0 - C^2 = 0.$$

This equation is equivalent to one which can be directly integrated by standard formulae:

$$\frac{d(\cosh\psi)}{\sqrt{\cosh^2\psi - \cosh^2\psi_0}} = \pm C_1 d\xi \quad \therefore$$

$$\cosh\psi = \cosh\psi_0 \cosh[C_1(\xi - \xi_0)] \tag{4.26}$$

where ξ_0 is another constant of integration, and we used the parity of hyperbolic cosine in order to get rid of the ambiguity of sign. Using this last result in equation (4.25) we get

$$d\theta = \frac{C^2 d\xi}{\cosh^2\psi_0 \cosh^2[C_1(\xi - \xi_0)] - 1}$$

which can be integrated using the example 2.458.2 from (Gradshteyn & Ryzhik, 2007), with the result

$$\tan\left(\frac{\theta - \theta_0}{C_1 \sinh\psi_0}\right) = \sinh\psi_0 \coth[C_1(\xi - \xi_0)] \tag{4.27}$$

θ_0 being still another integration constant. The equation (4.27) and the last equation of (4.26) provide our solution to differential system (4.24). It depends on four real parameters: θ_0, ξ_0, ψ_0 and C_1, which can be fixed by some boundary conditions, appropriate to our problem. This may turn out to be a routine, but it does not touch the essentials of the physical argument. However, that argument is strongly influenced by the fact that the vector **a** is completely arbitrary in the definition (4.23) of the parameter ξ.

Our issue is now obvious: it is represented by the difference in the results of these two variational problems. Even though the metrics are referring to the very same situation, they represent nevertheless two different physical objects. The second one is clear, for it is referring to a precise physical situation: an orbit varying continuously in its plane. Neither the eccentricity of this orbit, nor its orientation in the plane of motion are independent free parameters. As to the first of these variational problems, its physical objects are a little more complicated, but can nonetheless be briefly presented in the following manner. As known, within general relativity the gravitational field is expressed by the coefficients of the quadratic form representing the metric of spacetime continuum. The material content of this continuum is then assumed to control its physical properties through Einstein field equations. As it turns out, in special conditions of spacetime symmetry, the problem of solving these equations can also be reduced to a solution of the problem of harmonic mappings from the regular space to hyperbolic plane. This fact was brought to light by Frederick J. Ernst for *stationary* spacetime metrics, first in the case of cylindrical space symmetry (Ernst, 1968a,b) and then for a general situation not involving special space symmetry (Ernst, 1971).

Ernst himself (Ernst, 1968b) noticed the fact that a functional relation between the pure gravitational and pure electromagnetic potentials can solve the problem of gravitational field. In fact he proved (Ernst, 1971) that Einstein's field equations are amenable to the variational principle

$$\delta \iiint \left\{ R(\gamma) + 2\frac{\gamma^{mn}(\nabla_m \in)(\nabla_n \in^*)}{(\in + \in^*)^2} \right\} \sqrt{\det(\gamma)}\,(d^3\mathbf{x}) = 0$$

where $R(\gamma)$ is the scalar curvature of the three-dimensional metric γ, corresponding to a stationary four-dimensional relativistic metric, and $\in$ is the so-called *Ernst potential*. We can see here that in a flat space or, with some minor provisos, in a *space of constant Riemannian curvature*, this principle involves exclusively the complex Ernst potential. The details of expression of three-dimensional curvature tensor in terms of Ernst potential, and of expression of this potential in terms of gravitational and electromagnetic fields can be found in (Perjés, 1970), and especially (Israel & Wilson, 1972). A recent work (Mazilu & Porumbreanu, 2018) explains in detail, along this line of thought and with historical arguments, the implications of general relativity. The essentials of this line seem to emphasize the Ernst's idea that a relationship between gravitational and electromagnetic field is mandatory in solving the problem of metric structure of spacetime. To wit, Israel and Wilson (*loc. cit.*) show that in a flat three-dimensional environment, where the curvature tensor of the metric γ is null, there should be a linear relationship between the electromagnetic field and the Ernst potential. In this case, the scalar curvature $R(\gamma)$ is also zero, and therefore the spacetime stationary metric can be described exclusively in terms of the Ernst potential:

$$\delta \iiint \frac{\gamma^{mn}(\nabla_m \in)(\nabla_n \in^*)}{(\in + \in^*)^2}\,\det(\gamma)\,(d^3\mathbf{x}) = 0. \qquad (4.28)$$

In other words, only in cases where the gravitational field defines somehow an electromagnetic field, or vice versa, the *gravitational field is described exclusively by a complex potential.*

In order to better grasp the importance of such a circumstance, it pays to notice that the wave mechanics, in its quantum form, is liable to decide the issue, insofar as it concerns the planetary model of the fundamental structure of our model of the world. Indeed, as we already have noticed before, the quantum mechanics was built upon an approximation of the planetary model, which can adequately describe the world of electromagnetic phenomena at the microscopic level. On the other hand, a world described according to the prescription of general relativity is certainly at the other end of the scale of things scientific – the universe at large. And if the electromagnetic field determines a metric which is cosmological, then this very fact speaks of the identity of the two – *wave mechanics* and *general relativity* at *any scale transition* of the human knowledge. This simple observation is liable, by itself, to offer full support to Nottale's approach to SRT. However, there is more to it than meets the eyes.

For now, just notice that the description by Ernst potential is much more palpable if instead of the original Ernst potential $\in$ we use the field variable $h \equiv i \in$, so that the equation (4.28) becomes

$$\delta \iiint \frac{\gamma^{mn}(\nabla_m h)(\nabla_n h^*)}{(h - h^*)^2} \det(\gamma)\,(d^3\mathbf{x}) = 0. \qquad (4.29)$$

Obviously this variational equation describes a harmonic map between the ordinary flat space of metric tensor (γ_{mn}) and the complex half plane possessing the Poincaré metric, exactly as in the case of Kepler motion. And so it comes that this circumstance reveals here an essential possibility of interpretation related to the *physical idea of confinement*.

A Newtonian Brief on Density

Indeed, the situation above can be turned into a mathematical method of characterizing the matter in general, provided the space containind matter can be geometrically described by a metric having physical meaning. The previous discussion shows that a notorious place of the presence of matter, namely the nucleus of planetary atom, may be taken as a portion of space where the matter is continuous, just as the philosophical principles of general relativity demand. Then the theory of harmonic mappings can assume a general physical meaning: given a metric that characterizes the matter, *a harmonic map represents the mode in which the matter fills the space available to it*. In order to illustrate this statement we need, again, some reference to the previous example related to the nucleus of planetary model.

The classical mode in which the matter fills the space is by continuity, described through the Newtonian *concept of density*, which is established by comparison, via Archimedes law for fluids. We reproduce from Newton a significant excerpt involving the whole group of concepts which this method involves:

> COR. III. All spaces *are not equally full*; for if all spaces
> were equally full, then the specific gravity of the fluid which

fills the region of the air, on account of the extreme density of the matter, would fall nothing short of the specific gravity of quicksilver, or gold, or any other most dense body; and, therefore, neither gold nor any other body, could descend in air; for bodies do not descend in fluids, unless they are specifically heavier than the fluids. And if the *quantity of matter in a given space* can, by any rarefaction, be diminished, *what should hinder a diminution to infinity?*

COR. IV. If all the solid particles of all bodies are of the same density, *and cannot be rarefied without pores*, then a void, space, or *vacuum must be granted.* By bodies of the *same density* I mean those whose inertias are in proportion to their bulks [(Newton, 1974), Vol. II, p. 414; *our Italics*].

The very first thing to notice here is that the Archimedes' principle is referring to non-miscible bodies, i.e. to that situation characterized by Erwin Madelung as "penetrating each other without merging". This means that, properly generalized, this principle can be applied to matter at any spacetime scale.

The concept of density in the Newton's acceptance naturally enters here in explaining the "quantity of matter in a given space", by the definition that starts the *Principia*:

The *quantity of matter* is *the measure of the same*, arising from its density and *bulk* conjointly. [(Newton, 1974), Vol. I, p. 1; *our Italics*].

In the preceding metric description of the matter, it is however always possible to define locally, therefore exactly, the density as a relationship between the elementary volumes as parts of spaces represented by the metrics of space and matter. This is simply done as solution of a problem of mathematical principle, via a well known algorithm: *find the solution of the harmonic mapping problem, giving the coordinates in matter as functions of the coordinates in space, and then construct their Jacobian.* According to the principles of classical fluid theory, which certainly apply to the Madelung fluid, the density of matter should be proportional to the square of Jacobian of such a transformation. Sounds rather simple, but physically this approach raises a few challenging issues.

The key issue is an honest answer to that straight question of Newton: *what should hinder a diminution to infinity?* Obviously, Newton thought the 'nothing' for answer. He could not grasp that the answer is positive and just stares at him, due to the manner in which the matter presents itself to our senses: *the very confinement of matter.* This, however, became clear only in the last century, when the idea of confinement of matter took a dramatic turn of representation, which transformed it from an *a priori* thing offered to our senses, to a concept that needs theoretical explanation. This theoretical explanation starts nevertheless from the very same point where Newton himself started: *the telling of our senses*. Let us, therefore, get into some details.

The Concept of Confinement

Fact is that a metric can be always produced through the *mathematical* idea of confinement — and therefore in the general terms allowed by our senses for the idea of 'bulk' — by the Cayley's procedure of constructing metrics based on an absolute (Cayley, 1859): *a space region of any shape can always be considered as confined inside a closed space surface*. Usually, when constructing such a metric in geometry, one follows the original work of Arthur Cayley, and takes a quadric as the closed surface necessary to carry out the procedure. This, however, is not at all a mandatory choice. Dan Barbilian has shown that a Cayleyan metric can be constructed based on any algebraic surface represented by a homogeneous equation (Barbilian, 1937). He has noticed that if, in general, $Q(R_1, R_2, R_3)$ is a "ternary quantic" of degree 'n' — which, in modern algebraical language means a homogeneous polynomial of degree 'n' in three variables — then the quadratic differential

$$(ds)^2 = K^2 \left[\frac{d^2Q}{n(n-1)Q} - \left(\frac{dQ}{nQ} \right)^2 \right], \tag{4.30}$$

where K is a constant, generalizes the metric of Cayley, which is only a particular case of this for n = 2, as we just said. Here dQ is the first order differential, while d^2Q is the second order symmetric differential of the polynomial Q.

This mathematical construction qualitatively represents indeed the general property of matter as it first appears to our senses, of being confined in a finite volume. So, by Cayleyan idea of metric, the intuitive property of being confined to a volume can be directly turned into a concept that includes the very modern theoretical idea of confinement of the material components of the nucleus. For instance, in the case of eccentricity of a Kepler orbit, if the coordinates R_k are the components of the eccentricity vector in any Cartesian reference frame in space, we can consider the cubic $Q(R_1, R_2, R_3) \equiv R_1 \cdot R_2 \cdot R_3$. This ternary cubic may be taken as representing the volume of a cuboid having the edges R_1, R_2 and R_3, which contains the matter of the atomic nucleus. Based on this volume, the quadratic differential form (4.30) can be written as

$$ds^2 = -\frac{K^2}{18} \left[\left(d \ln \frac{R_2}{R_3} \right)^2 + \left(d \ln \frac{R_3}{R_1} \right)^2 + \left(d \ln \frac{R_1}{R_2} \right)^2 \right]. \tag{4.31}$$

This metric has quite a few physical connotations that can apply to the theoretical concept of matter in general, of which this time only two are of interest

71

in order to accomplish our task here. These are, however, in close physical correlation with each other, and that correlation is what interests us most.

First, recall that the Hubble parameter is defined as the logarithmic derivative of a linear size of the universe, with respect to the cosmic time. If the "shape of the universe" is a tensor, $\mathbf{R}$ say, having the eigenvalues R_k, then its logarithmic derivative has the eigenvalues $\ln(R_k)$ and when referred to a cosmic time, the metric (4.31) is *Misner's variance of the Hubble parameter*, which can be written as (Misner, 1968):

$$3(\Delta H)^2 = \left(\frac{\dot{R}_2}{R_2} - \frac{\dot{R}_3}{R_3} \right)^2 + \left(\frac{\dot{R}_3}{R_3} - \frac{\dot{R}_1}{R_1} \right)^2 + \left(\frac{\dot{R}_1}{R_1} - \frac{\dot{R}_2}{R_2} \right)^2. \quad (4.32)$$

This kinematical parameter is therefore an expression of the existence of a metric characterizing the size of the universe based on a measure of its volume, and can be presented as an absolute metric in the manner above. Of course, this characterization is not unique by any means, forasmuch as the measure of the volume is not unique, but this can be delegated to a precise mathematical study. For, two things need to be noticed here, both of them in connection with the *concept of matter* as it lies at the foundations of general relativity.

The first one of these, concerns the idea of extension of the meaning of volumic measure: the cubic Q represents the measure of *extension in space of the matter in general,* just as in the case of the central matter in the classical Kepler problem. This clearly defines a reference frame, however not of the usual Euclidean type. One should realize indeed, that the quantity $(R_1 \cdot R_2 \cdot R_3)$ which, as we just said, *is physically a volume*, remains invariant with respect to any permutation of the indices of symbols R_1, R_2, R_3. In that case we can construct *three different vectors* obtained by even permutations from one another, but having the same Euclidean length, and therefore *providing the same eccentricity of the orbit*:

$$\mathbf{e}_1 = \begin{pmatrix} R_1 \\ R_2 \\ R_3 \end{pmatrix} ; \mathbf{e}_2 = \begin{pmatrix} R_3 \\ R_1 \\ R_2 \end{pmatrix} ; \mathbf{e}_3 = \begin{pmatrix} R_2 \\ R_3 \\ R_1 \end{pmatrix}. \quad (4.33)$$

These vectors also have the same Euclidean angle between them, defined by the equation

$$\cos \psi = \frac{\mathbf{e}_i \cdot \mathbf{e}_j}{\sqrt{(\mathbf{e}_i \cdot \mathbf{e}_i)(\mathbf{e}_j \cdot \mathbf{e}_j)}} = \frac{\sum R_2 R_3}{\sum R_1^2} \quad (4.34)$$

where the symbolic sums run over the positive permutations of the symbols. As, in the generic case, $R_1 \neq R_2 \neq R_3 \neq R_1$, these vectors are linearly independent, and they can be taken as establishing a frame, however not always an orthonormal one. The angle ψ is a distinctive feature of such a frame, and cannot be $\pi/2$

if all the three quantities R_k are positive. Therefore, a shape statistics (Nedeff, Lazăr, Agop, Eva, Ochiuz, Dumitriu, Vrăjitoriu & Popa, 2015), for instance, of the volume of matter thus described, can never define an orthonormal frame this way. However, such a frame is perfectly legitimate, and certainly qualifies as an *affine reference frame*. There is, in this respect, a noteworthy case where this reference frame cannot be but exclusively orthogonal. For this we have to take the liberty offered to us through the possibility of constructing a metric by any homogeneous cubic, no matter of its geometrical meaning.

In order to explain what we mean by this, it is worth noticing that if the symbols R_1, R_2 , and R_3 represent the roots of a cubic equation in general, for instance, the eigenvalues of a matrix, then the columns (4.33) *establish a reference frame uniquely related to these eigenvalues*. On the other hand, the eigenvectors of the matrix establish, in turn, another reference frame describing *in space* the physical quantities embodied in the matrix representation. If the matrix is symmetric, then this last reference frame is orthonormal. For the case in which the matrix is a tensor, its eigenvalues can be presented as what we would like to call *Novozhilov averages* (Mazilu & Agop, 2012) of the physical magnitudes represented by tensors. These are *space averages over directions in space*, performed in a orthonormal reference frame given by the eigenvectors of the symmetric tensor at hand:

$$\overline{m_n} = \Omega^{-1} \oiint_{\hat{\mathbf{u}} \cdot \hat{\mathbf{u}} = 1} m_n d\Omega_{\hat{\mathbf{u}}}; \quad \overline{m_t^2} = \Omega^{-1} \oiint_{\hat{\mathbf{u}} \cdot \hat{\mathbf{u}} = 1} m_t^2 d\Omega_{\hat{\mathbf{u}}}. \quad (4.35)$$

Here the generic tensor is denoted by letter $\mathbf{m}$— suggesting... magnitude. The averages are done over directions represented on the unit sphere represented by the unit vector $\hat{\mathbf{u}}$. Ω is the whole solid angle of the space around a given position (4π, to be precise), while $d\Omega_{\hat{\mathbf{u}}}$ is the elementary solid angle centered around $\hat{\mathbf{u}}$. The original case, addressed by Valentin Valentinovich Novozhilov, was the one of tensions and deformations (Novozhilov, 1952) representable, as well known, by symmetric 3×3 Euclidean tensors. There are, in general, only two Novozhilov averages that can be calculated this way, and these are referring to a *normal component* of the tensor, which in our notations here would be m_n, and a *tangential component* of the tensor, which in our notation would be m_t . The labels 'normal' and 'tangential' are referring to an arbitrary plane in space, for a tensor quantity can be projected in two ways on a plane of normal $\hat{\mathbf{u}}$: first along the normal, giving the 'normal' quantity:

$$m_n = \sum_k m_k u_k^2$$

and secondly in the plane giving the 'tangential' quantity:

$$m_t^2 = \sum_{k=1}^{3} m_k^2 u_k^2 - \left(\sum_{k=1}^{3} m_k u_k^2 \right)^2$$

73

via Pythagoras' theorem. Then the averages in equation (4.35) embody a fact of continuity: the measurement of the tensor in a certain point reveals averages over a *sphere of arbitrary small radius*, inside of which any *orientations of planes are a priori equally probable*. These measured quantities cannot be but only functions of the position in which they are measured. Their expressions are:

$$3\overline{m_n} = \sum m_1; \quad 15\overline{m_t^2} = \sum (m_2 - m_3)^2 \qquad (4.36)$$

where the sign of sum indicates this time a sum over all positive permutations of the numerical indices. In cases where these two expressions are the only available as results of the measurement, the eigenvalues of **m** are reproducible up to an arbitrary angle, which we like to call the *Barbilian phase* (Barbilian, 1938). The recipe is the following (Mazilu & Agop, 2012): construct the complex quantity $h = u + iv$, where i is the imaginary unit, with $u = \overline{m_n}$ and $v = [(5/6)m_t^2]^{1/2}$. Then we have, up to a permutation of the indices of eigenvalues:

$$\begin{aligned}
m_1 &= u + v\tan\phi, \\
m_2 &= u + v\tan(\phi + 2\pi/3), \qquad (4.37) \\
m_3 &= u + v\tan(\phi + 4\pi/3)
\end{aligned}$$

where ϕ is an arbitrary angle of orientation of the vector representing tangential component of the tensor in one of the *eight octahedral planes* from the position of measurement. Thus, we have here a genuine 'eightfold way' which can be described based on a simply transitive continuous group having the SL(2,R) structure, in the variables u, v and ϕ, with three parameters: the *Barbilian group* (Barbilian, 1938).

A Clasic Example of Affine Reference Frame: Maxwell Stress Tensor

Now, after this short presentation, let us describe our promised case of a mandatory orthogonal affine reference frame generated by the eigenvalues of a tensor. Any physical conclusion involving the light is based upon local measurements. Because a measurement always involves some kind of uncontrolled averages over regions of space and and sequences of time, this may be taken as being the case with the local measurements of an electromagnetic field. And, to the extent in which an electromagnetic field, can be defined by the *Maxwell stress tensor*, we can simply apply the previous theory to Maxwell stresses. As known (Stratton, 1941), this is a tensor **M** of the form ('M' here stands for 'Maxwell')

$$\mathbf{M} = \lambda\mathbf{e} + \mu\mathbf{b} \quad \therefore \quad m_{ij} = \lambda e_{ij} + \mu b_{ij} \qquad (4.38)$$

74

where the matrices **e** and **b** are obtained from the vectors of intensity of electric field and induction of the magnetic field, which will be denoted by the same bold letters, via the relations

$$e_{ij} = e_i e_j - (1/2)\mathbf{e}^2 \delta_{ij} \quad \mathbf{e}^2 = e_1^2 + e_2^2 + e_3^2,$$
$$b_{ij} = b_i b_j - (1/2)\mathbf{b}^2 \delta_{ij} \quad \mathbf{b}^2 = b_1^2 + b_2^2 + b_3^2.$$

The eigenvalues of Maxwell stress tensor (4.38) are:

$$m_1 = m, \quad m_{2,3} = \pm\sqrt{m^2 - \mathbf{p}^2} \tag{4.39}$$

where we have used the notations

$$m \equiv (1/2)(\lambda \mathbf{e}^2 + \mu \mathbf{b}^2); \quad \mathbf{p} = \sqrt{\lambda\mu}(\mathbf{e} \times \mathbf{b}). \tag{4.40}$$

This pair even represents an eigenvalue of **M** and its associated eigenvector – *the Poynting vector*. The other two eigenvectors are in the plane given by the two original vectors **e** and **b**. Now, if we calculate $\cos\psi$ from equation (4.34) using the eigenvalues (4.39), the result is

$$\cos\psi = \frac{\mathbf{p}^2 - m^2}{2\mathbf{p}^2 - 3m^2}. \tag{4.41}$$

Thus the affine reference frame established according to the recipe (4.33) from the eigenvalues of the Maxwell tensor, can be orthogonal only in the singular case where $|\mathbf{p}| = \pm m$, so that the geometrical angle θ between the two vectors **e** and **b** in their plane is given by

$$\sin\theta = \frac{\lambda e^2 + \mu b^2}{2eb\sqrt{\lambda\mu}}. \tag{4.42}$$

The quantity in the right hand side here is always ≥ 1. Therefore it cannot represent the sine of a proper geometrical angle but only in the case when it is 1. But in that case $\theta = \pi/2$, and this is always the case of the *electromagnetic field in vacuum*. Therefore, inasmuch as the previous physical theory is based on electromagnetic light, for instance, the affine reference frame it selects is always an orthogonal Euclidean frame, just as the reference frame given by the eigenvectors of the Maxwell tensor.

The light, however, has also color, and in astrophysics, for instance, the color makes all the difference in extracting the information about the distant objects. It is then quite remarkable that a second connotation of the metric from equation (4.31) is related to the *geometrization of the manifold of colors*. This time, however, the geometry reveals its general meaning as a necessary tool giving the possibility of *arithmetization* of a manifold. We take here the concept of arithmetization as the general possibility of *attaching numbers as measures to physical properties*, based on the general theory of ensembles. The spirit of this approach is best grasped, in our opinion, by Nicholas Georgescu-Roegen, who connected the probabilities with an *ordering of scales of the world things* (Georgescu-Roegen, 1971).

Chapter 5. The Light in a Schrödinger Apprenticeship

One can hardly say that in the year 1920 Erwin Schrödinger was still an apprentice in the field of physics. Experimental works, a few of them accompanied by exquisite theoretical explanations and conclusions, produced by Schrödinger for a decade since his graduation, can easily testify to the contrary. And yet, judging from the perspective of the epoch-making idea of wave function introduced by him six years later, we venture to advocate the idea that the year 1920 was one of his 'internship' so to speak, in theoretical physics. For, that year shows up as being the year when Schrödinger rounded up some incongruent pieces of a concept, important at that historical moment, just as it is today for that matter, mostly for practical industrial and medical purposes: *the concept of color*. And, in our views, the fact that starting from 1920 Schrödinger dedicated his life exclusively to theoretical physics, carries a fundamental meaning especially related to this circumstance: *the necessity to arithmetize a continuum*. In view of one of the observations of Charles Galton Darwin, reproduced here in a previous excerpt, which we take as a definition of the concept, an *interpretation in physics* must be related to such a fundamental procedure. Not only this, but as Nottale's approach to the transition of scale suggests, the procedure itself needs to be inserted explicitly as part and parcel of the theoretical physical description.

The concept of color fully illustrates the necessities of this procedure, but not only that: in hindsight it clears up the Newtonian idea of matter, by properly framing the *Archimedes' principle* of measurement of densities. For, the difference between the quantification of densities and the quantification of colors is, nevertheless, a major one: in the case of colors the randomness is dominant, as a modern relic of the old exclusive physiological approach of the calibration and comparison of colors. This is not to say that nowadays the physiology does not play any part in the definition of the colors, because it does, mostly for its own benefit, but also for the benefit of theoretical physics, in general. What we want to emphasize is the fact that the subjectivism here is mainly related to the statistical side of the color issue, and the metric approach of the color continuum, to the geometry of which Erwin Schrödinger has fundamental contributions, may have contributed substantially to the ideas which resulted in the creation of the concept of wave function as we have it today. Thus, if Louis de Broglie has the widely recognized merit of participating, and almost solitary at that, to the creation and development of the objective side of the duality wave-particle, one can surely say that Erwin Schrödinger has such a merit but referring to the 'subjective' side of that concept. The recognition of this fact comes from the specific participation of Schrödinger to rounding up the modern concept of color, and thus it takes us unexpectedly to the very foundations of the modern theoretical physics.

At the time of Schrödinger's contribution on the theory of colors, this one has already have gotten out of the realm of pure experimental work assisted by an occasional mathematics, entering the phase of physiological character based on a firm mathematical principle, through the works of Hermann von Helmholtz. That firm mathematical principle was the theory of metric spaces. In his famous habilitation dissertation from 1854, which set the ground for the modern theory of metric manifolds, Bernhard Riemann wrote an often-cited phrase:

> The positions of the *objects of sense*, and *the colors*,
> are probably the only familiar things whose specifications
> constitute multiply extended manifolds (*our Italics*).

This phrase has been taken as a mathematical research incentive ever since its publishing thirteen years later (Riemann, 1867), mostly in geometry, and mostly regarding the first "thing" of the two, viz. "positions of the objects of sense", whereby the "specifications" are mainly made by reducing the spatial extension of the objects to points in a three-dimensional manifold. The colors are, however, "things" through the agency of which those "objects" are made available to our senses, but they did not seem to get so much attention, at least from a mathematical point of view. The work of Hermann Günther Grassmann on a linear theory of color, almost contemporary with that of Riemann (Grassmann, 1853), had not received so much attention, just like Riemann's work for that matter, and in fact one can assume that the whole life of Grassmann stands witness of such a kind of social rejection. It was not so much the inexactness of Grassmann's laws regarding the mathematics of color continuum, which the physiology started to assess only lately (Pokorny, Smith & Xu, 2012), that contributed to this state of the matters, as much as it was the absence of social means for undersanding their true drive.

The rival theory of colors, created by Hermann von Helmholtz (Helmholtz, 1852), had a substantially better chance which, in hindsight, can be explained by the fact that it was predisposed to a metric description (Helmholtz, 1891, 1892). And in this instance, from the very same algebraic point of view, it requires linearity only at a certain level, which best fits the color measurement needs. In fact one can say that insofar as the practice of mathematics is concerned, the theory of Helmholtz is closer to the Riemann's original views.

However, it has the great merit of notifying the human spirit on the big difference between the "extended manifolds" of "positions of the objects of sense" and the "extended manifolds" of the human means of sensing those objects: the interpretation is referring here to ensembles having the *cardinality of continua*. For, as Schrödinger expressed it:

> The *manifold of lights* has a higher power than the *power of the continuum*, namely that of a space of functions; and hence an indefinitely large number of dimensions. A priori it would seem possible that this could also hold true for the *manifold of color qualities*, or at least that it could have had a very large number of dimensions; as in the case of manifold of combined tones, since the ear acts to some extent as a harmonic analyzer. *That is not the case here*. Rather, according to *the principle of matching appearance on adjacent fields*, the lights arrange themselves into large groups — each one of the power of the function space — and the manifold of these groups of the same appearance is, for normal color-perception persons, of the dimension three the highest ever observed. This dimensionality *is a fundamental fact of basic colorimetry*, and its derivation from experience will be our closer concern here. [(Schrödinger, 1920), *our translation and Italics*; see also (Niall, 2017), *Ch. II, first communication*]

Obviously, there is a difference between the *manifold of lights* and the *manifold of color qualities*. In other words, the color, considered as a property of light, is more than frequency, which is its basic physical association even nowadays. The main point here is that the eye is not simply a harmonic analyzer, at least not to the same extent as the ear, for instance. Fact is that the light is produced independently of a material structure accessible to our senses, while the sound depends essentially on such a structure. This very fact, i.e. the accessibility to our senses, seems to make an essential difference. Anticipating a little some results of our discussion here, we can say that the color is an expression of transcendence from *transfinite* to *finite* order, once it can be discovered in the relationship between the frequency and the wavelength.

But, let us start from the very beginning: the above excerpt contains the essentials of what has driven Schrödinger's contributions to the color theory, and in fact his main contributions to the future wave mechanics, soon to have its first milestone set by Louis de Broglie from a purely physical point of view (de Broglie, 1923). Those essentials are obviously abridged here by the idea of *cardinality of continua*, necessary in case one thinks to give a continuum the simplest structure necessary for interpretation: that of an ensemble. In retrospect, one can say that Schrödinger was compelled to take this action by the unsatisfactory concept of the very same physical light ray contemplated six years later by Louis de Broglie, which *did not contain properly the feature of color*. This fact is not quite so explicitly set by Schrödinger from the very beginning, but gets to us through the explanations he offers in the works on color from 1920. In order to make the reasons of his contribution more obvious, it is perhaps better to reveal that contribution from the very beginning. And we cannot accomplish such a task without first getting into description of the influence upon this subject of another pillar of the modern natural philosophy, which was Hermann von Helmholtz, known for his fundamental contributions to many different chapters of the physics of 19th century.

A Special Contribution of Helmholtz

The color is, obviously, one of the most important among the physical properties of the objects revealed to us by senses, and is usually described as a continuous three-dimensional manifold. The reader interested on every minute detail of measurement procedures and theoretical account of light, may consult any edition the comprehensive work of Günther Wyszecki and Walter Stanley Stiles on *Color Science*. We had at our disposal (Wyszecki & Stiles, 1982). By and large the mathematical theory of color stays in the shadow of the quote we presented above from the celebrated Dissertation of Bernhard Riemann from 1854, "Über die Hypothesen, welche der Geometrie zu Grunde liegen', which established the grounds of the modern metric geometry. However, we are interested here in another idea of Riemann, much more significant in our context. It touches the difference of two procedures that have always been mixed, sometimes even inadvertently, with each other in calculations of the density of matter, as we have exemplified previously with an excerpt from Feynman. To wit:

> *Parts of a manifold* distinguished by a characteristic or a boundary are called *quanta*. Their comparison regarding quantity takes place in case of *discrete extensions by counting*, in case of *continuous extensions by measurement*. The measuring consists in a *superposition of the extensions to be compared*; for measuring, therefore, a means is required to carry *one extension as a meter* for the other. If this is lacking, one can only compare two extensions if one is a part of the other, and even then only to decide for the more or less, not for how much. The investigations which may be made of them in this case represent a general part of the doctrine of extensions, independent of measures, where the extensions are not regarded as independent of location, and not as expressible by a unity, *but as domains in a manifold*. For several

parts of mathematics, especially for the treatment of multi-variate analytic functions, such investigations have become a necessity, and their absence is probably a major cause that the famous Abel theorem and the achievements of Lagrange, Pfaff, Jacobi for the general theory of differential equations were unfruitful for so long. For the present purpose it suffices to emphasize two points of this general part of the doctrine of the extended quantities, where nothing else is assumed other than what is already contained in the concept of quantity, of which the first is the *generation of the concept of a multiply extended manifold*, and the second concerns *the reducing of the positional determinations in a given manifold to quantitative determinations*, thus making clear the essential characteristic of an n-fold extension. [(Riemann, 1867); *our translation and Italics*]

With Hermann von Helmholtz, the measurement of colors takes a fundamental turn prompted by necessities of physiology of vision. Specifically, one needed a means of quantitatively describing that 'comparison of the extensions' mentioned by Riemann. At this juncture, the idea of metric was taken by Helmholtz as a measure of the difference in the "extensions to be compared". This is part of the general philosophy which says that in order to declare things identical, one needs first to know their difference. And the physiology of vision already had such a case of measuring the difference in things by the difference in sensations: it was the *law of Fechner*, which stipulated that the intensity of sensation is proportional to the logarithm of stimulus engendering that sensation within the eye.

So, Helmholtz was the first to construct a significant metric for the manifold of colors (Helmholtz, 1891), based on Fechner's law of relation stimulus-sensation. In our reference frame above, R_1, R_2, R_3 would represent the *primary colors*, nowadays usually red, green and blue, at least as far as routine eye vision is concerned. Then any color can be linearly represented, indeed, with respect to these primaries by the quantity:

$$R = x^1 R_1 + x^2 R_2 + x^3 R_3 \qquad (5.1)$$

where x^k are the amounts of corresponding primary color R_k contained in the quantity representing the color R. Mantaining the same primaries, which is the customary case in the modern theory of colors, defines a 'universal reference frame' as in the regular Euclidean geometry. *In this reference frame*, a certain color can be therefore represented by the 'ket':

$$|x\rangle = \begin{pmatrix} x^1 \\ x^2 \\ x^3 \end{pmatrix}. \qquad (5.2)$$

Now, Helmholtz's problem was to construct a Riemannian metric for these quantities in the reference frame given by the three fixed primaries. His procedure can be best categorized in modern geometrical terms by the statement that there is a certain colorimetric approach of establishing the amounts x^k of the primaries, whereby the metric can be taken as the difference between the colors distinguished by a normal eye. Quoting:

Determination of similar color pairs. If we have a pair of composite colors, one of which contains the *quanta* x, y, z, *of the primary colors*, the others which differ a little (x + dx), (y + dy), (z + dz), and the *intensity of the first colors can be increased* in the ratio $1 : (1 + \varepsilon)$, so that their components become

$$x(1 + \varepsilon) \quad , \quad y(1 + \varepsilon) \quad , \quad z(1 + \varepsilon)$$

then the measure of the *difference in feeling* between the second and third of these colors is

$$\frac{dE^2}{k^2} = \left(\frac{dx - \varepsilon x}{x} \right)^2 + \left(\frac{dy - \varepsilon y}{y} \right)^2 + \left(\frac{dz - \varepsilon z}{z} \right)^2.$$

With variable values of ε, *this becomes a minimum* when

$$0 = -2 \left(\frac{dx - \varepsilon x}{x} + \frac{dy - \varepsilon y}{y} + \frac{dz - \varepsilon z}{z} \right)$$

or

$$\frac{dx}{x} + \frac{dy}{y} + \frac{dz}{z} = 3\varepsilon. \tag{3a}$$

Now, if we give to ε this value, we get the value of the minimum of dE^2 as

$$\frac{dE^2}{k^2} = \frac{1}{3} \left\{ \left(\frac{dx}{x} - \frac{dy}{y} \right)^2 + \left(\frac{dy}{y} - \frac{dz}{z} \right)^2 + \left(\frac{dz}{z} - \frac{dx}{x} \right)^2 \right\} \tag{3b}$$

$$\frac{dE}{k} = \frac{\delta\lambda}{\sqrt{3}}$$

$$\times \sqrt{ \left(\frac{1}{x}\frac{dx}{d\lambda} - \frac{1}{y}\frac{dy}{d\lambda} \right)^2 + \left(\frac{1}{y}\frac{dy}{d\lambda} - \frac{1}{z}\frac{dz}{d\lambda} \right)^2 + \left(\frac{1}{z}\frac{dz}{d\lambda} - \frac{1}{x}\frac{dx}{d\lambda} \right)^2 }$$

The x, y, z which are used to indicate the color value of the various spectral colors of Messrs. A. König and C. Dieterici, are related to the elementary colors R, G, V, according to Newton's law by linear homogeneous equations, but their coefficients are initially unknown. Let us denote these values

$$\left.\begin{array}{rl} x = & a_1 \cdot R + b_1 \cdot G + c_1 \cdot V \\ y = & a_2 \cdot R + b_2 \cdot G + c_2 \cdot V \quad \cdots\cdots\cdots \\ z = & a_3 \cdot R + b_3 \cdot G + c_3 \cdot V \end{array}\right\}. \qquad (4)$$

It should first be noted that each of the coefficients in each horizontal row can be given an arbitrary value, since

$$\frac{dx}{x}, \quad \frac{dy}{y} \quad \text{and} \quad \frac{dz}{z}$$

do not change their values if each of the amounts x, y, z is modified by an arbitrary constant factor. In other words, the choice of the coefficients in the sense of Young's theory is only limited by the fact that, the values of R, G, V belonging to the spectral colors, no negative values may be given to x, y, z. This will never be the case if all the coefficients a, b, c have positive values. [(Helmholtz, 1891b); *our translation and Italics; original captions*]

The problem of definition of the colorimetric terms is hardly settled even today, but hopefully the Italicized phrases in this excerpt are self-explanatory to some point, at least for the geometrical purposes which we follow closely in the present work. The reader interested in details has at his/her disposal, besides the already cited comprehensive work of Wyszecki and Stiles, a compendium by David L. MacAdam containing excerpts from principal inspirational works in colorimetry along the time, rendered in a historical order (MacAdam, 1970). A work of Deane B. Judd, closer to our times is particularly interesting, inasmuch as it contains definitions of colorimetric terms, even with translations from German (Judd, 1940); as a matter of fact any glossary published by modern optical societies will do.

Now, Helmholtz's definition of the 'minimum difference in feeling' has a precise geometrical meaning, which completely justifies his mathematical procedure. Certainly, his metric (3b) characteristic to the situation of such a minimum, is no different from (4.32), and this may have an explanation in the fact that the Pogson-type correlation, used for associating the distance to the magnitude of the astrophysical objects, is in a sense similar to Fechner's physiological law. At the time of von Helmholtz, this law has been held as essential in the comparison and quantification of the color continuum by direct vision. The excerpt above illustrates, in the first place, the way he thought to construct a global metric of this color continuum: geometrically, in modern terms, this is *a metric possessing a special Killing vector.* Indeed, if such a vector is taken as a displacement in the color continuum, the quadratic form of the metric is invariant with respect to this displacement if (Vrânceanu, 1962):

$$\xi^k \frac{\partial g_{ij}}{\partial x^k} + g_{il} \frac{\partial \xi^l}{\partial x^j} + \frac{\partial \xi^l}{\partial x^i} g_{lj} = 0; \quad x^i \to x^1 + \xi^i \qquad (5.3)$$

where **g** is the metric tensor. Now, it is clear that the displacements sought for by Helmholtz are given by vectors of the form $\xi^k = \varepsilon \cdot x^k$, so that the Killing equation is of the form:

$$x^k \frac{\partial g_{ij}}{\partial x^k} + 2g_{ij} = 0. \qquad (5.4)$$

Therefore, all of the entries of the metric tensor *must be homogeneous functions of degree -2 in the coordinates ordering this continuum.* The choice of Helmholtz is

$$\frac{dE^2}{k^2} = \left(\frac{dx^1}{x^1} \right)^2 + \left(\frac{dx^2}{x^2} \right)^2 + \left(\frac{dx^3}{x^3} \right)^2 \qquad (5.5)$$

and obviously satisfies the Fechner's law in the sense that the differential components of the metric do not change if the amounts of primaries are "modified by an arbitrary constant factor". So the minimum of the metric should formally coincide with the Misner's parameter of anisotropy of the universe, and this may not be just a formal coincidence, as long as the Hubble parameter is evaluated by light measurements concerning the magnitude of astrophysical objects. However, two main objections can be raised here

Enters Erwin Schrödinger

The first one of these was laid down by Erwin Schrödinger, and concerns, on one hand, the parameter to which the Fechner's law refers, in case it is indeed a law, and on the other hand, the correctness of that law. He noticed that the theory of color continuum should be centric affine, and that the *surface of equal brightness* 'h' should be a surface orthogonal to the direction of position vector. According to his logic, the function $h(\mathbf{x})$ has to satisfy the condition:

$$h(\mathbf{x}) = x^k \frac{\partial h(\mathbf{x})}{\partial x^k}, \quad \frac{\partial \ln h(\mathbf{x})}{\partial x^i} = \frac{g_{ij} x^j}{g_{kl} x^k x^l} \qquad (5.6)$$

which the Helmholtz's choice does not satisfy. Indeed, for the metric (5.5) the second condition here is

$$\frac{\partial \ln h(\mathbf{x})}{\partial x^i} = \frac{1}{3x^i} \quad \therefore \quad h(\mathbf{x}) = const \cdot \sqrt[3]{x^1 x^2 x^3} \qquad (5.7)$$

and h($\mathbf{x}$) *is not a linear function* as the first condition (5.6) requires. However, Schrödinger is convinced that the experimental data up to date (i.e. year 1920) shows that the brightness is linear to a very good approximation, a condition not so sound, as it was shown in later experiments. Be it as it may, the bottom line of the Schrödinger's analysis was a proposal for a little more general metric tensor of the color continuum. His metric tensor is still diagonal, but has as components more general entries, still homogeneous of degree -2, involving three arbitrary constants [(Schrödinger, 1920), " *the Part III of communication,* equation (4.12)]. Schrödinger's conclusion was that having such a metric...

> ... we obtain constant intervals following **Fechner**'s Law
> not only for the same stimulus quality but also *over the whole*
> *gamut of color, as experience appears to require* over a large
> region of color space. [*loc. cit. ante, our Italics; translation*
> *from* (Niall, 2017)]

This conclusion touches the second one of the objections we have in mind here, which gives to Schrödinger's intervention a genuine character from the point of view of the future physics.

The 'experience' here is that physics leading to quantum theory, which requires essentially 'the whole gamut of color'. This is, indeed, the fundamental difference between Fechner's and Pogson's laws: the first one is refering to a precise stimulus (color), while the second one is referring to a finite portion of spectrum, if not even at the whole spectrum. From the point of view of this experience, the color measurements should satisfy the principles referring to the *blackbody radiation,* and thus there is physics to address such measurements. As a matter of fact this physics existed as such even at the time Schrödinger developed his theory. For, as one can notice right away, Helmholtz's way of acceptance of the Fechner's law, expressed in the special form of the Killing vector of the metric, is actually the mathematical basis of demonstration of Wien's displacement law, and therefore the general mathematical premise of Planck's law of radiation. This last law refers to the *spectral density of radiation* and even gives the known expression for it. Although this expression is a consequence of a specific statistical behavior of the radiation (Mazilu & Porumbreanu, 2018), its statistical standing — as a connection between ensemble average and variance function — forbids us from taking it as a density. But it *can be taken nevertheless as a probability density*. Specifically, it is a Gaussian density of probability for *the cubic root of the ratio between frequency and temperature* (Priest, 1919).

This conclusion sounds rather speculative, even hazardous we might say, for the state of mind of a physicist used to the classroom physics of the last part of the previous century, whereby the Planck's law is a kind of taboo, so to speak. This is why we did not just take for granted the old arguments of Irwin G Priest himself, but documented this conclusion based on many sets of classical data on blackbody radiation (Mazilu, 2010), and also on the modern official data on cosmic bacground radiation (Fixsen, Cheng, Gales, Mather, Shafer, & Wright, 1996). In this last case numbers are at the public's disposal, and using those numbers the conclusion formulated above seems to be satifactorily true, at least just as satisfactory as the official conclusion that the cosmic background radiation is a Planck-type blackbody radiation. Therefore the Planck's law can be taken as a probability law, as long as it can be reduced to such a fundamental law by a transformation used currently in the statistical practice (Box & Cox, 1964). However, the implications of this conclusions are much more profound, for they target the very transition between the microcosmos and the universe in the case of a SRT. This fact seems rather intuitive if we think of the data on blackbody radiation: it characterized the atomic world, and all of a sudden proves to be valid at the scale of the universe.

However, two things are forced now upon us now, and require our close consideration. First, the theoretical background of the demonstration of Planck's in the microscopic world is the physical adiabatic transformation. The history proved that the adiabatic transformation needs an appropriate geometry from the perspective of wave theory related to nonstationary Schrödinger equation. Secondly, we may need a reformulation of the Wien's displacement law, insofar as a probability density equivalent to Planck's law — like that of Priest — does not seem to contain a factor free of temperature. Both these issues have been addressed in a moment of the history of physics related to the name of Sir Michael Victor Berry.

Chapter 6. The Wave Theory of Geometric Phase

To Madelung, just like to Schrödinger himself, the stationary Schrödinger equation was an eigenvalue equation. Likewise, the nonstationary equation was worth considering just as long as it provided this eigenvalue equation. In the beginning, though, the connection was not quite as direct as it seems today. To wit, the stationary equation was instrumental in guiding the reason towards a nonstationary equation, for both, Madelung as well as Schrödinger himself. The main difference in the two approaches is that Madelung assumes a Born-type of interpretation for the wave function in connection with density of a fluid, and this, just like the actual probabilistic interpretation of the wave function, leaves the phase of the wave function undecided. In view of the identification of the phase with the classical action, this lack of decision might not be quite out of line: after all, the classical action itself was always undecided in theoretical mechanics. However, the wave mechanics, with its quantum amendment, brings an important point to fore: a specific intervention of the geometry, in fact, of the differential geometry.

Enters Sir Michael Berry

It is in this matter indeed, that the main result of Sir Michael Berry's work on the geometry required by the existence of arbitrary phase factors (Berry, 1984) makes its mark, showing that the nonstationary Schrödinger equation means much more than usually thought, when taken from the perspective of fundamental knowledge. To wit, based on nonstationary Schrödinger equation, Berry noticed that the *adiabatic changes* of the phase in the definition of the wave function ask for a definite differential geometry of the adiabatic parameters. Thinking back, one must admit that the nature of such a geometry, as well as that of the associated statistics, was somehow obscured in the initial works on the interpretation of quantum mechanics, as this interpretation started from considerations on *a single* adiabatic parameter only, and remained concentrated upon this case for a long while (Mazilu & Porumbreanu, 2018). With the work of Berry, though, the adiabatic evolution, in general, gets the explicit position of *a transition between two scales of spacetime,* much in the way Erwin Madelung pointed it out (Madelung, 1927). For once, this points to a position of the wave function close to, if not identical with, that conferred to it by Nottale's SRT. On the other hand, however, it is more important to notice that Berry's work actually gives a precise place to the *potential function* in the geometry of the world in general. Let us therefore pinpoint the essentials of this theory, by following the line of thought of Michael Berry.

The time dependent Schrödinger equation to be considered here is:

$$i\hbar|\dot\Psi\rangle = H(\mathbf{R},\mathbf{r})|\Psi\rangle. \tag{6.1}$$

The notation suggests the Born–Oppenheimer approximation, according to which in describing the behavior of the 'fast' variables $(\mathbf{p},\mathbf{r})$ – in this special instance the phase space position parameters of the point of reduced mass μ – there are 'slow' variables $(\mathbf{P},\mathbf{R})$, that can be taken as *adiabatic parameters* – as are, for the very same instance the coordinates of the position of the total mass M (Born & Oppenheimer, 1927). In other words, the geometry of adiabatic parameters is here a geometry of the space proper. On the other hand, it should be admitted that the wave function must have several components, a suggestion which explains the representation of the wave function above as a 'ket': $|\Psi\rangle$. The Hamiltonian controlling the motion of the fast variables is

$$h(\mathbf{R},\mathbf{r}) = \frac{\mathbf{p}^2}{2\mu} + V(\mathbf{R},\mathbf{r}) \tag{6.2}$$

depending on the slow position through potential $V(\mathbf{R},\mathbf{r})$. The ensuing stationary Schrödinger equation (Darwin, 1927):

$$h(\mathbf{R},\mathbf{r})|\mathbf{R}\rangle = \varepsilon(\mathbf{R})|\mathbf{R}\rangle \tag{6.3}$$

then provides the eigenfunctions and the corresponding eigenvalues of the problem, which, of course, will parametrically depend on the slow position. This dependence is what brings forth the *Berry geometrical connection* in the space of slow variables and this connection takes a physical shape via a *vector potential* defined as

$$\mathbf{A}(\mathbf{R}) \equiv \langle\mathbf{R}|i\hbar\nabla_{\mathbf{R}}|\mathbf{R}\rangle. \tag{6.4}$$

Indeed, in his 1984 work Michael Berry took note of the fact that if, according to the prescription of the adiabatic approximation, one has to write the solution of nonstationary Schrödinger equation corresponding to the Hamiltonian (6.2) in the form

$$|\psi(t)\rangle = \exp\left(-\frac{i}{\hbar}\int_0^1 ds \cdot \varepsilon(\mathbf{R}(s))\right) e^{i\gamma(t)}|\mathbf{R}(t)\rangle \tag{6.5}$$

then the arbitrary phase $\gamma(t)$ *is not integrable in a cyclic transformation* in the space of adiabatic parameters given by the slow position. More to the point, using the nonstationary Schrödinger equation, the phase $\gamma(t)$ proves to be a solution of the ordinary differential equation

$$\dot{\gamma}(t) = i\mathbf{A}(\mathbf{R}(t)) \cdot \dot{\mathbf{R}}(t) \tag{6.6}$$

with $\mathbf{A}(\mathbf{R})$ defined by (6.4), so that in a cyclic transformation along a closed curve in the space of slow parameters, the wave function gains a phase factor *depending on the cycle*:

$$e^{i\gamma(C)}; \quad \gamma(C) = i \oint_C \mathbf{A}(\mathbf{R}) \cdot d\mathbf{R}. \tag{6.7}$$

Now, what is more important for our discussion at this juncture, is a fact which Berry himself noticed: if the fast variables $(\mathbf{p}, \mathbf{r})$ are the spin variables represented by a vector $\mathbf{S}$, then the Hamiltonian

$$h(\mathbf{R}, \mathbf{S}) = \mathbf{S} \cdot \mathbf{R} \tag{6.8}$$

is invariant with respect to the rotations of the adiabatic spatial background. So he found that *in this case*, the *curvature associated to the connection* (6.4) is simply the magnitude of intensity of a magnetic field given by a well-known expression

$$\mathbf{B}(\mathbf{R}) = g\frac{\mathbf{R}}{R^3}. \tag{6.9}$$

Notice, however, that insofar as it should be referred to as a curvature, this vectorial expression is in fact representative for many other fundamental physical situations. The chief among these is the one involving the classical Newtonian central force responsible for gravity.

In fact, as Barry Simon recognized in a work conferring universality to Berry phase (Simon, 1983), a trait which, we should say, is intuitively appropriate for a fundamental ingredient of the natural philosophy of any taste, the vector (6.9) itself is universal and its presence in the theory is only a matter of the language necessary for adapting the theory. Quoting:

> Since Berry is talking about integrating *(e,de) (i.e. a differential 1-form, a.n.)* along curves which he makes analogous to a vector potential, *he talks about magnetic monopoles.* Since we only care about *(de,de) (i.e. a differential 2-form, a.n.)* whose dual is divergenceless away from degeneracies, *we do not use the magnetic monopole language.* Since the dual may not have a zero curl, *electrostatic language is not appropriate.* Since the sources have a sign, we *still use the phrase*

88

charge for the coefficient of the delta function in d[(de,de)]
at singularities. [(Simon, 1983), Note 15; *our italics*]

The universality conveyed to Berry's phase by the work of Barry Simon, needs to be somehow explained in the most general terms, and we shall do this here with reference to a *three- dimensional phase space*: the usual positions with respect to a space reference frame must be, in our opinion, considered as 'phases' of the classical material points. As we shall presently show, this explanation retains the statistical spirit of the original discovery descending directly from the early stages of the two new mechanics, quantum and undulatory [(Berry, 1985); (Hannay, 1985)]. However, our approach has the advantage of avoiding the classical Hamiltonian formalism, which demands a precise knowledge of the energetical content of a system. This, in our opinion, is still wanting in physics as, in fact, it was always wanting (Poincaré, 1897).

Again, we find it beneficial discussing the general principles based on a well-known example, and this is, naturally, the so-called Aharonov-Bohm effect (Aharonov & Bohm, 1959). This case comes in handy here, because it is referring to the nonintegrability of the phase factors representing the indecision in the definition of the wave function in closed space loops, which, as the work of Berry clearly reveals, should have a universal connotation. Thus, considering the adiabatic parameters' space as a configuration space necessary in the process of interpretation after all, the wave function must be a state function! — we impose upon the evolution in this space the Liouville theorem, according to which the elementary volume of the space should be preserved along the motion. In these conditions the motion can be described by a field of vectors, and this field of vectors decides its time *as a succession of time moments*, helping to order the set of states along the motion. In spite of the manner in which it is customarily taken in theoretical physics, the time is never a matter of free choice, and in this respect a Liouville dynamics is just about the only one of its kind, set on satisfying a correct definition of time. The Figure 1 from (Berry, 1984) for instance, should be related to such an interpretation.

We describe the dynamics, in some general coordinates, x^1, x^2, x^3 say, obtained with the benefit of a reference frame. Assume further that with respect to these coordinates the oriented elementary Liouville volume of the space is given by the exterior differential 3-form:

$$\Omega = dx^1 \wedge dx^2 \wedge dx^3. \qquad (6.10)$$

Then, according to Vladislav Nikolaevich Dumachev, this dynamics has the following geometrical grounds (Dumachev, 2009, 2010, 2011). Generally, as in any phase space having a physical meaning, the exterior differential form of the elementary volume is preserved through the flow given by a vector **X** if its Lie derivative along this vector vanishes. This condition allows us to assume that the time derivative of a differential form is given by the Lie derivative along a path having the time as a natural continuity parameter. Postponing, for the moment, the detailed explanation of this last concept, let us concentrate upon the consequences of such identification. For an exterior differential form in general, the Lie derivative is calculated via Cartan's 'golden rule', viz.:

$$L_{\mathbf{X}}(\Omega) = i_{\mathbf{X}}(d \wedge \Omega) + d \wedge i_{\mathbf{X}}(\Omega) \qquad (6.11)$$

where $i_{\mathbf{X}}(\Omega)$ is the projection of the exterior form Ω along the vector $\mathbf{X}$ [see (Arnold, 1976) for a detailed explanation of the concept, and significant examples]. If the space is three-dimensional and the differential form is (6.10), then $d \wedge \Omega = 0$ because Ω is a 3-form, and then the vanishing of Lie derivative equation reduces to

$$\frac{d}{dt}\Omega \equiv L_{\mathbf{X}}(\Omega) = 0 \quad \therefore \quad d \wedge i_{\mathbf{X}}(\Omega) = 0. \qquad (6.12)$$

By the Poincaré lemma [(Spivak, 1995), pp. 94ff] this means that, in fairly general conditions, the exterior form $i_{\mathbf{X}}(\Omega)$ is exact, so that there is a differential form, 'h' say, allowing us to write:

$$i_{\mathbf{X}}(\Omega) \equiv \Theta = d \wedge h. \qquad (6.13)$$

Obviously, here we cannot have but two different cases, depending on the degree of the exterior differential form 'h', which in turn depends upon geometrical nature of the vector field $\mathbf{X}$. Specifically:

(1) if $\mathbf{X}$ is a vector proper, i.e. in the coordinate basis it is expressed as

$$\mathbf{X} \equiv X_1 \partial_1 + X_2 \partial_2 + X_3 \partial_3 \qquad (6.14)$$

then the differential form Θ, as defined by equation (6.13) is 2-form, inasmuch as for $i_{\mathbf{X}}(\Omega)$ we have the expression:

$$i_{\mathbf{X}}(\Omega) = X^1 dx^2 \wedge dx^3 + X^2 dx^3 \wedge dx^1 + X^3 dx^1 \wedge dx^2. \qquad (6.15)$$

Thus, 'h' must be a 1-form:

$$h \equiv h_1 dx^1 + h_2 dx^2 + h_3 dx^3 \qquad (6.16)$$

while from equation (6.13) we deduce that the vector $\mathbf{X}$ must be assumed a curl from the mathematical point of view, i.e. for theoretical necessities:

$$X_h^1 = \partial_3 h_2 - \partial_2 h_3; \quad X_h^2 = \partial_1 h_3 - \partial_3 h_1; \quad X_h^3 = \partial_2 h_1 - \partial_1 h_2. \quad (6.17)$$

90

This vortex is generated by the differential form 'h' from (6.16), which explains the lower index in our notation of the components of the vector $\mathbf{X}$.

(2) if $\mathbf{X}$ is bivector — equivalent in this particular case with a skew-symmetric matrix — which in the coordinate basis can be written as:

$$\mathbf{X} \equiv X_1 \partial_2 \wedge \partial_3 + X_2 \partial_3 \wedge \partial_1 + X_3 \partial_1 \wedge \partial_2 \qquad (6.18)$$

the differential form Θ defined in equation (6.13), is a 1-form, so that 'h' must be a 0-form, i.e. a regular function. In this case the vector $\mathbf{X}$ must be assumed a gradient for theoretical necessities, having the components

$$X_1^h = \partial_1 h; \quad X_2^h = \partial_2 h; \quad X_3^h = \partial_3 h \qquad (6.19)$$

with an obvious meaning for the upper index of the components. Thus, in the classical theory of gravitation for instance, one can assert that the volume element is carried along the force vector. However, if the force is a gradient it is carried on a surface by a vortex, while if the force is a curl, it is carried along a line. This is the well-known duality revealed by the Maxwell's electromagnetic theory, which turns out to be quite a general rule.

A Kepler Motion Analysis: the Geometrical Condition of Yang-Mills Fields

Our quintessential example in indicating the procedure of constructing the phase space is, again, the dynamical solution of the classical Kepler problem. Taken as such, this problem highlights indeed a fundamental element of physics upon which we need to insist especially: the *quantitative definition of time by the physical process of revolution of the celestial bodies*. This definition is stipulated in the second of the Kepler laws. Here we shall extend the problem of this time, but from the above mathematical point of view, by speculating on the fact that in a two- dimensional world, the elementary area 2-form is in fact the Liouville elementary volume, the two-dimensional analogous of the 3-form from three-dimensional case represented by equation (6.10).

Thus the plane of the the classical dynamics associated with a Kepler motion can be organized as a phase plane in a classical sense, where, nevertheless, the momenta corresponding to coordinates are not necessary. Indeed, the coordinates themselves are reciprocally momenta to one another in a natural way, based on the differential 1-form of area. If 'x' and 'y' are the generic coordinates of this 'phase plane', then exactly as in the three-dimensional case, from the Lie derivative (6.11) only the second term survives:

$$L_{\mathbf{v}}(dx \wedge dy) = d \wedge i_{\mathbf{v}}(dx \wedge dy) \qquad (6.20)$$

91

because in a two-dimensional world the exterior differential of a 2-form vanishes. Thus, the rate of variation of the 'Liouville elementary volume' should be given here by

$$\frac{d}{dt} dx \wedge dy = d \wedge i_{\mathbf{v}}(dx \wedge dy) = d \wedge (v_x dy - v_y dx) \tag{6.21}$$

where $\mathbf{v}$ is a certain velocity field. If this elementary area is preserved by transport, then the differential 1-form in the right hand side of this equation should be exact, so that by Poincaré lemma, there is a 'current function', ψ say, such that

$$v_y = -\frac{\partial \psi}{\partial x}; \quad v_x = \frac{\partial \psi}{\partial y}. \tag{6.22}$$

Thus, the velocity field $\mathbf{v}$ satisfies to partial differential equation:

$$\frac{\partial v_x}{\partial x} + \frac{\partial v_y}{\partial y} = 0. \tag{6.23}$$

In cases where a potential exists as well, both this potential and the current functions are solutions of the Laplace equation:

$$\mathbf{v} = \nabla \phi \quad \therefore \quad \nabla^2 \phi = \nabla^2 \psi = 0. \tag{6.24}$$

Now, if the velocity field is linear and homogeneous in the coordinates of this motion:

$$\begin{pmatrix} v_x \\ v_y \end{pmatrix} = \begin{pmatrix} A & B \\ C & D \end{pmatrix} \cdot \begin{pmatrix} x \\ y \end{pmatrix} \tag{6.25}$$

the condition (6.23) is satisfied only when the matrix there is traceless: A + D = 0. Consequently if the evolution is Hamiltonian, in the transformation (6.25) only three parameters from the four ones are essential. This could be seen even directly: the 1-form in the left hand side of the equation (6.21) can be written as

$$v_x dy - v_y dx = \frac{1}{2}d[-Cx^2 + (A - D)xy + By^2] + \frac{A + D}{2}(ydx - xdy) \tag{6.26}$$

so that, given the condition (6.23) it is an exact differential. In this case, redefining the parameters by the following identifications:

$$-C \equiv \alpha; \quad A \equiv \beta; \quad B \equiv \gamma$$

the equation (6.26) can be written in the form

$$v_x dy - v_y dx = \frac{1}{2}d(\alpha x^2 + 2\beta xy + \gamma y^2). \tag{6.27}$$

92

One can thus read directly the form of the function required by the condition of vanishing of the elementary area rate during the transport: *it is the quadratic form giving the Keplerian trajectory as a conic.* The equation (6.25) can be written as

$$\begin{pmatrix} v_x \\ v_y \end{pmatrix} = \begin{pmatrix} \beta & \gamma \\ -\alpha & -\beta \end{pmatrix} \cdot \begin{pmatrix} x \\ y \end{pmatrix}. \qquad (6.28)$$

This tells us that the velocities linear in coordinates, necessary for vanishing of the elementary rate of area, as given by equation (6.27), are tangent to the family of conics

$$\alpha x^2 + 2\beta xy + \gamma y^2 = Q \qquad (6.29)$$

where Q is a constant; they represent trajectories having the same center.

An observation of general interest in the theoretical characterization of the matter is worth extracting from the developments up to this point. Due to equation (6.21), the vector $\mathbf{v}$ qualifies as a Killing vector field. Assuming linearity, as in equation (6.25), this Killing vector is of the *Helmholtz type from the theory of color vision*, only more general, satisfying the observations of Schrödinger, as it were. Until a proper occasion to make use of this observation, let us only notice the meaning of the linear relation (6.25): it can be taken *as defining the coordinates of a classical material point* having the velocity $\mathbf{v}$. Case in point: the definition (4.7) of the eccentricity vector, which is related to the position of *a classical material point in matter.* However, let us continue with the mathematical development here, to see where it leads us.

The potential function ϕ does not exist here but only in the cases where the 1-form

$$v_x dx + v_y dy \equiv (\beta x + \gamma y)dx - (\alpha x + \beta y)dy \qquad (6.30)$$

is an exact differential. In the most general conditions in parameters — i.e. in the cases where α, β, γ *are different*, even arbitrary with respect to each other — the motion along the trajectory (6.29) under condition (6.30) must satisfy to a special constraint — *different from the vanishing of exterior differential of the 1-form* (6.30) — showing that it is an exact differential. The condition of vanishing in (6.30) would only mean a restriction in the parameters α, β, γ; however, proceeding as before, i.e. putting the equation (6.30) in the form exhibiting an exact differential:

$$v_x dx + v_y dy \equiv \frac{1}{2}d[\beta x^2 + (\gamma - \alpha)xy - \beta y^2] + \left(\frac{\gamma + \alpha}{2}\right)(ydx - xdy) \quad (6.31)$$

we can give directly the potential without any reference to the parameters. Indeed, the restrictive case in parameters: $\alpha + \gamma = 0$, a consequence of the equation $d \wedge (v_x dx + v_y dy) = 0$, is obviously included in this transcription of the 1-form. Aside from this case — very special indeed, when it comes to the general parametrization of the orbit with respect to its center — from equation

(6.31) follows that only if the motion is such that the rate of the area swept by the position vector with respect to the center of the conic (6.29) is a constant – the *area constant* from the Kepler problem, which we shall denote from now on by the symbol "å" – a potential can exist, which is however not uniquely defined, but depends upon the area constant and time. Indeed, if:

$$ydx - xdy = \dot{a}dt \qquad (6.32)$$

then (6.31) becomes

$$v_x dx + v_y dy \equiv d\phi(\mathbf{x}, t) = \frac{1}{2} d[\beta x^2 + (\gamma - \alpha)xy - \beta y^2] + \left(\frac{\gamma + \alpha}{2}\right)\dot{a}dt$$
$$(6.33)$$

which means that the velocity potential is defined up to an arbitrary additive constant by

$$\phi(\mathbf{x}, t) = \left(\frac{\gamma + \alpha}{2}\right)\dot{a}t + \frac{1}{2}[\beta x^2 + (\gamma - \alpha)xy - \beta y^2]. \qquad (6.34)$$

This equation represents therefore the condition that the motion thus defined is performed along the normal to the trajectory (6.29).

Obviously, *in order not to constrain the parameters* of the Keplerian trajectory of a classical material point, we ought to pay a price. In the latter case here this comes down to vanishing of the differential form $(v_x dx + v_y dy)$. From a geometric point of view, this only means, indeed, that the velocity field thus defined is perpendicular to the trajectory. However, with reference to such trajectories, i.e. objects geometrically located by the three parameters α, β, γ, the vanishing of our 1-form means a *deformation described by the gradient of a potential*, 'u' say, in a form rendering the true meaning of this velocity field:

$$\begin{pmatrix} v_x \\ v_y \end{pmatrix} = \begin{pmatrix} 0 & 1 \\ -1 & 0 \end{pmatrix} \begin{pmatrix} u_x \\ u_y \end{pmatrix} ;$$
$$\begin{pmatrix} dx \\ dy \end{pmatrix} \equiv dt \begin{pmatrix} u_x \\ u_y \end{pmatrix} = dt \begin{pmatrix} \alpha x + \beta y \\ \beta x + \gamma y \end{pmatrix} .$$
$$(6.35)$$

In other words, the condition that the area rate can be defined the way it is defined in mechanics (i.e. by the position vector and its time rate), is in fact given by the *existence of a deformation of the Keplerian orbit*, admitting a linear field of rates of deformation.

Before anything else should be said here, let us notice that the area rate (6.32) is calculated with reference to the center of the orbit. Should this rate be calculated with respect to the focus, this equation would then get the form

$$ydx - xdy = a \cdot e \cdot d(x - y) + \dot{a} \cdot dt \qquad (6.36)$$

94

where 'e' is the eccentricity of the orbit, and 'a' is its major semiaxis. Now, this raises an important problem: recall that the center of force in the classical Kepler problem is actually in the focus of orbit, while its center − even though qualifying as a center of force according to Newtonian theory of forces − is just *a space position with no physical content* from the point of view of the classical natural philosophy. However, Newton's way of defining the forces which he invented for describing the gravitation, is by a *genuine theory of measurement* (*Principia*, Corollary 3 of the Proposition VII from Book I): the ratio of any two forces pulling simultaneously a certain point on the actual orbit should be recognized in a specific way in the parameters of the orbit. Now, in the Kepler problem one of these two forces certainly pulls toward focus, for that is the position where the attractive matter is located. However, it is only when the other point is taken as the center of the orbit that one obtains the central force of magnitude inversely proportional with the square of the distance to focus. Thus, the second force involved in the Newtonian procedure of describing the central forces is generated by... nothing, for in the center of the orbit there is no matter at all! But this force has an entirely different reality, made gradually obvious in physical optics especially after the works of Augustin Fresnel. We shall come to this issue again later on.

For the moment, in order to describe the deformation above in terms of the variation of orbit, as physically seems just a natural thing to do, notice that we can define the variation of a quadratic form like that from equation (6.29) by the usual rule of differentiation. This means that the variation in question can be written in the form

$$\delta Q = \langle x|d\boldsymbol{\alpha}|x\rangle + \langle dx|\boldsymbol{\alpha}|x\rangle + \langle x|\boldsymbol{\alpha}|dx\rangle; \quad \boldsymbol{\alpha} \equiv \begin{pmatrix} \alpha & \beta \\ \beta & \gamma \end{pmatrix}. \qquad (6.37)$$

In so writing this variation, we intended to draw attention upon the fact that, when defining the variation this way, we have two differential contributions: one coming out of the variation of the matrix of the quadratic form, the other one resulting from the variation of the position vector itself in the plane of motion. We are now interested in somehow separating these variations, in order to introduce the physics through the entries of the matrix α. For instance, in the classical Kepler problem where the quadratic form is coming out from the second principle of dynamics for a Newtonian force, these entries have the special form from equation (4.8), depending on the *initial velocity* chosen for the motion. In general, therefore, physically speaking, the parameters of the orbit are a matter of choice, actually even a convenient choice for the problem at hand. Therefore, we would be interested in separating the two variations through some *universal conditions*, *i.e. conditions independent of the actual position of the revolving material point*, and therefore imposed exclusively upon the matrix from equation (6.37). One of these conditions seems obvious here:

95

we just need to require that the variation of orbit should be due exclusively to the variation of the matrix. Using (6.37), this condition can be translated into:

$$\delta Q = \langle x | d\boldsymbol{\alpha} | x \rangle \quad \therefore \quad \langle dx | \boldsymbol{\alpha} | x \rangle + \langle x | \boldsymbol{\alpha} | dx \rangle = 0. \qquad (6.38)$$

Thus, in order that the variation of the orbit should not depend but on the variation of its defining matrix, it is sufficient that the bilinear form $\langle dx | \boldsymbol{\alpha} | x \rangle$ vanishes. This defines a connection between the differentials of the position coordinates and the differentials of the matrix entries, which can be expressed as:

$$\begin{pmatrix} dx \\ dy \end{pmatrix} = (dt) \begin{pmatrix} \beta & \gamma \\ -\alpha & -\beta \end{pmatrix} \cdot \begin{pmatrix} x \\ y \end{pmatrix} \qquad (6.39)$$

where 'dt' is, for the moment, a differential of some arbitrary 'time' parameter. A comparison between (6.28) and (6.39) shows that this last equation can be interpreted as an equation of the 'motion' along which the 1-form (6.27) must vanish, while the quadratic form from that equation is constant, thus providing a conservation law for that motion. Accordingly, the equation (6.39) defines the velocity field we contemplate here. On the other hand, if that velocity field is thus defined, the variation of the quadratic form (6.37) *cannot be accomplished but only through the variation of its parameters*. This enforces the general conclusion: if the motion takes place along a classical Keplerian orbit, then the variation of that orbit cannot be accomplished but only by the variation of its parameters.

The Berry Moment of Human Knowledge

We just stated above that this last universal condition is obvious, but this remark might not be quite so obvious for the casual observer. Fact is, however, that the conclusion here does not tell us anything new over what we already know from the history of physics. Actually, the conclusion represents the fundamental thesis of Niels Bohr's quantization procedure, in its utmost generality: in the classical Kepler description of the planetary atom problem we have to do with fixed orbits, and the transition among these can be accomplished only by jumps from an orbit to another. As the physical parameters of the orbit are the only 'coordinates' of these geometrical objects, then such a jump ought to be characterized by a variation of the physical parameters. Now, in the modern quantum description of the planetary atom, the procedure of quantization was inspired by thermodynamics. This is why the variation in question had to be described *adiabatically*, leading to the theory of Heisenberg, and further on to the idea of operator associated to a physical quantity.

From this perspective one can say that there is a 'Berry moment' of the human knowledge, which carries, among others, the precise meaning of liberating the knowledge from the lock-up of adiabaticity, and this is reflected in the general attributes of this landmark of physics. Indeed, what we just called the 'Berry moment' has essentially three major conceptual contributions related to the name of Sir Michael Berry. The first one, and from gnoseological point of view perhaps the most important among them, is the one just described succinctly above (Berry, 1984), showing that the adiabaticity must actually be allotted to geometry. Many people think that this work is indeed the most important of Berry's works in the problem we discuss now, and we are not here set on disagreeing. Rather, we just need to emphasize the fact that the work from 1984 is only a partial conclusion of what seems to be a constant concern of Michael Berry. And this concern can be understood as such, only by making a 'moment' of knowledge out of it, which would be a proper moment of knowledge only if we consider at least two other works of essence of Michael Berry, both related to issues regarding the idea of adiabaticity. To reveal those issues we need a little digression on a significant point of the physical knowledge, which changed the world view in the last century. That significant point is represented by the implementation, in fact, as well as only in idea, of what we think is the essential ingredient of our knowledge: *the physical reference frame*.

Here again, an excerpt from the quintessential work of Charles Galton Darwin is in order, helping to express the connection between the idea of reference frame and the space, in its right context: the *concept of interpretation*. To wit, *we need to make a concept* out of the procedure of interpretation. In a word, the active knowledge is not possible without the objective duality of de Broglie type:

> In dealing with the *interpretation* we have touched on one of the great difficulties which have made it hard to gain physical insight into the wave theory. This is the fact that *the wave equation is not in ordinary space but in a co-ordinate space*, and the question arises *how this co-ordinate space is to be transcribed into ordinary space*. It would appear that most of the difficulty has arisen from an attempt *to apply it illegitimately to enclosed systems*, which *are really outside the idea of space*. In most of the problems we shall discuss, the question hardly arises, but where it does the correct procedure is so obvious that there is no need to deal with it in advance. *It is tempting to believe that this will be found to be always the case.* [(Darwin, 1927); *our Italics*]

The development of physics invalidated this last 'tempting belief' mentioned by Darwin. More to the point, the only place to apply it *legitimately* appears to be exactly where "most of the difficulty" is concentrated: *to enclosed systems*. In our opinion this defines both the necessity of an "ordinary space" and the right "interpretation" of waves in terms of particles. The key point is that a reference frame helping in constructing a "coordinate space" is, *actually, always an enclosure.*

The quintessential physical reference frame remains forever the Earth, of course: it is the reference frame carrying us through the universe, the reference proper of all our positive knowledge. However, the first physical realization of a reference frame needed in the measurement as understood by Darwin in his work, but having at least part of the properties of Earth, is due to Wily Wien and Otto Lummer (Wien & Lummer, 1895), and is referring to what is generally known in physics and natural philosophy as a 'hohlraum'. The fact that we attach this experimental tool to the physics of reference frame is warranted by some significant circumstances suggesting indeed the idea of a *reference frame.* Of chief importance among these is the *blackbody radiation*: the apparatus of Wien and Lummer was actually intended for the study of this physical system. However, it is well known that the incentives of such a study came from astrophysics, and led physics to the discovery of quantization. Based on this discovery, the relativistic cosmology made a 'hohlraum' out of the entire universe, and this image was confirmed by the outstanding discovery of the cosmic background radiation (Penzias & Wilson, 1965), which is a blackbody radiation corresponding to an equilibrium temperature of approximately 2.7K. As we just said, these facts are actually well known.

However, rarely, if ever, is it explicitly stated that the conclusion they entail — namely that the astrophysical background radiation is a proof of the behavior of universe we live in — is a consequence of a *warranted extension* of the conclusions drawn from the laboratory measurements on blackbody radiation. And that warranty comes from the fact that the *Wien displacement law*, which is the condition that every law of radiation must satisfy, is a consequence of the invariance with respect to the dimensions of the 'hohlraum', therefore a consequence of a *space scale invariance.* In other words, extending a little bit this conclusion, if a property of the blackbody radiation is discovered in a laboratory, it is true also in the universe, which is actually only a... workshop of the largest size among many others. The discovery of the expansion of the universe allowed speculations go even further, to the conclusion that the essential property of background radiation is a consequence of the expanding of the universe, which is actually a physical process representative for a certain physical structure of the universe (Dicke, Peebles, Roll, & Wilkinson, 1965). The whole demonstration of this fact had to be thermodynamical, in view of the uncertainties of the structure thus described, but this raised an entirely different problem, brought about by the quantum theory of radiation *per se.*

Namely, the *physical theoretical* demonstration of the right law of thermal radiation is based upon the existence of the adiabatic invariants, and in this case the expansion of universe ought to be an adiabatic expansion from the very thermodynamical point of view. While it may reasonably appear as such, that expansion is nevertheless controversial, as it was in fact from the very beginning. Not only that its rate remains undecided even today, being a kind of 'unavoidable hypothesis' believed as missing only a temporary precise quantitative confirmation. After all, this might not be a cause for denying the reality of the physical process, because the appearance of adiabaticity can be a relative phenomenon as far as its rate is involved. However, there are data showing that the universe is downright contracting, not expanding. This, again, might not be a theoretical problem: the expansion is a consequence of the metric description of the universe in spacetime and in some physical conditions, represented by the mean density of the matter in universe, the contraction might occur. An oscillating universe is just as acceptable for theorists as a purely expanding universe, but this cannot be a case of expansion existing simultaneously with contraction. Unless, of course, the universe behaves differently with the direction of its observation, i.e. it is utterly anisotropic to start with, as far as its matter content is to be considered. Be the case as it may, two problems arise and these bring us back to what we have called the 'Berry moment' of our knowledge.

First, the physics in a reference frame *defined as an enclosure*, should be independent of the variation of its dimensions, *no matter of its rate*. This fact — which is true for the Wien's displacement law — could not be proved explicitly in its universality, however it is proved for a theory that was always taken as the first physical cosmology: *the Newtonian cosmology*. In this cosmology the fundamental forces are taken to be central forces of magnitude inversely proportional with the square of distance between locations, and these are the only forces selected by the variation of the dimensions of the reference frame containing matter (Berry & Klein, 1984). In other words, the condition of adiabaticity taken *sensu stricto* selects a too narrow range of physical processes in describing the state of the universe, to start with. In fact we have just about only the cosmic background radiation in that range. However the first ever theoretical physical cosmology, the Newtonian cosmology, was not even considered from the point of view of the expansion of the universe, for if it would have been considered from such a stance, it would have been able to reveal that the Newtonian forces are universal forces in a precise sense, shown in the work of Berry and Klein just cited. So these forces carry at least the same gnoseological significance as the blackbody radiation, if not a more important one at all, which descends from Newton himself.

Indeed, everything in the 'Berry moment' has started, in this particular matter, from a discovery regarding the *collisions*: in studying a spherical enclosure containing classical material points — the classical model of ideal gas serving, for instance, in the definition of the absolute temperature — Klein and Mulholland have found an important invariant of collision between the material points and the enclosure wall, with respect to changes of the dimensions of the enclosure (Klein & Mulholland, 1978). The work of Berry and Klein that we cited above, analyzes, with the nonstationary Schrödinger equation, a general theory leading to this invariant, and finds that no matter what forces are involved in the coresponding Hamiltonian, there is always a *time scale* and a *length scale* where the forces appear to be conservative, provided the potential is updated by an additive harmonic oscillator potential (Berry & Klein, 1984). The scale transformation considered by Berry and Klein turns out to leave invariant the Newtonian inverse square forces. After all, this might have been the reason why these forces were so important in physics: Newton used the idea of collision in order to prove *the physical acceptability of his invention* [(Newton, 1974), *beginning of the second section of the* Book I].

The second aspect involved in the implementation of the physical reference frame as an enclosure, and represented in what we just have called the 'Berry moment' of knowledge, is the equivalence principle of the general relativity. In fact, it is about the very idea of 'Einstein's elevator', as one came a long time ago to call an essential physical ingredient of a well-known 'gedanken' experiment, destined to prove the equivalence between inertia and gravitation. Of course, some might consider that such an experiment is no more... 'gedanken' at all, in view of the present-day astronautical experience, but this is not the issue here. The issue is that one needs to recognize that such an elevator is *naturally equivalent to a Wien-Lummer experimental 'hohlraum'*, and should thus be considered on equal footing with the original one, at least in some fundamental theoretical aspects, if not in each and every detail of the theoretical problems of physics. For, with this issue we enter the essence of the very idea of interpretation in the sense of Darwin, namely the theory of fluids, and therefore, implicitly, the concept of a Madelung fluid.

For the first time, this fact was only tacitly suggested in the thermodynamical model of Enrico Fermi for the decaying of elementary particles (Fermi, 1950). Three years later, in 1953, it was updated by Lev Davidovich Landau, who invoked some *relativistic hydrodynamical motivations*, in order to account for the experimentally observed 'jet' phenomenon in description of the decay of particles [see (Landau, 1965), for a reproduction of the original work]. The proper quantum-mechanical description is still wanting, but we think that the attention should be focused on the fact that the jet phenomenon is typically related to the cosmic radiation showers. The work of Berry and Balazs on the *wave mechanics of the Airy wave pachets* (Berry & Balazs, 1979), allows us to say the following. An elementary physical particle can be considered as an Einstein elevator within the space of which the matter can be described hydrodynamically, by relativistic methods. In its downward motion toward Earth, this elevator reaches a point where the fluid, implicitly present in this hydrodynamic description, 'couples' somehow with the gravitational field, and around that event (moment and position) the internal 'hohlraum' behaves like a Madelung 'swarm' of free material points just like an ideal gas: it is the moment of *asymptotic freedom* in the 'hohlraum'. Specifically, there is a position within the hohlraum moving at constant (gravitational) acceleration, and with respect to that position each material point of the swarm behaves statistically like a harmonic oscillator. All this story remains to be later documented in the present work.

Let us stop here for a quick recap, in order not to loose the continuity of ideas. Because the initial quantization procedure was inspired by thermodynamics, the variation in the dimensions of a reference frame had to be adiabatic, thus properly warranting the Madelung assertion that we do not have here a ≪ "jump", but rather a slow transition into a state of non-stationarity ≫. In the case of hydrogen atom this condition led to the theory of Heisenberg, and further to the idea of operator associated to a physical quantity. However, the 'Berry moment' of our present knowledge would indicate that the rate of variation of the parameters of the orbit is completely indifferent, provided this variation is described in a certain association with the time of motion. A further conclusion then imposes almost by itself: the quantized electromagnetic fields from the prescriptions of the Bohr quantization, must be theoretically covered by the variation of the parameters of the orbit of a Keplerian system. In other words, the theoretical description of the electromagnetic field resides in the geometry of those parameters: *the electromagnetic fields are, in general, Yang-Mills fields to be described in a geometry of the manifold of positions of Kepler orbits of classical material points.* That geometry is a geometry of $\mathbf{sl}(2,R)$ Riemannian space, to be documented in what follows right away.

All of the previous conclusions can be simply summarized by the statement that the Keplerian motion, as classically described, entails with necessity the idea of quantization. It is indeed possible to talk about a jump between orbits only if these are real, i.e. when motions proceed along them. In such a case the jump cannot be described but by the variation of the parameters of an orbit. We ought to emphasize once again the fact that in the transition between classical and quantum mechanics one cannot talk about 'quantum revolution' as a sudden leap of of manifest continuity in our knowledge. It is the variation of the parameters of the orbit which is then physically represented by a field, and insofar as at a microscopic level this field is an electromagnetic field − generally a Yang-Mills field − the transition between orbits should be equivalent with the existence of such a field: this is exactly the content of the second postulate of Niels Bohr.

A Classical Implementation of the Idea of Interpretation

We can generalize this idea in the framework of classical natural philosophy, by a methodology of interpretation due to Heinrich Hertz, who conceived the classical material point — a *material particle* in his context — as a fundamental agent of physics (Hertz, 2003). Namely, according to Hertz, a *material particle* is that physical agent allowing us *to associate a certain position in space at a moment of time, with any other position in space at any other moment of time.* In other words, this definition actually states explicitly the property of the classical material point of being the physical agent allowing us to correlate any two different events. This property is only implicitly understood in the concept of classical material point and, up to the moment of rising of special relativity, taken as implicitly understandable in this concept. To Hertz, a *material point* proper is then *an aggregate of material particles.* Certainly, such definitions are the only ones suitable for a spacetime description of the universe of events. Once accepted, a first conclusion following from them is that the modern concept of confinement is just a natural limitation of the very possibilities of human knowledge. All it takes for such a conclusion is only a need of physical explanation of the natural property revealed to us by senses: *the matter does not exist but in spatially extended structures.* Since such a structure contains space inside it, the matter should be defined just by the opposite property: it is the place where *space does not have access.* Kepler problem reveals the way to conceive it, so we follow this way, different from that followed by Hertz himself in erecting his mechanics (Hertz, 2003).

First, notice that in the planetary atom an electron can be considered a material point in the sense of Hertz, i.e. a collection of material particles. These are the equivalents of de Broglie's photons in an elliptical capillary tube around nucleus, as we have shown before. Likewise, the atomic nucleus can also be *interpreted* as a Hertz material point. Then the Newtonian dynamics of every material particle from the matter of electron can be described by a Kepler orbit around a material particle inside the nucleus. Therefore an equation like (6.25), or (6.28) in fact, is actually a gauging, which gives positions in a de Broglie tube in terms of the velocities. There is a kind of continuity here that needs a *stochastic explanation,* because the 'swarm of material particles' of electron is always confined inside the de Broglie elliptic capillary tube. As each particle of electron can be paired by force in the Newtonian sense with each particle of the nucleus, it is then obvious that this pairing should proceed way faster than the motion of the electron around nucleus. The idea may be able to give a concept to the Madelung's observation about "the slow transition into a state of non-stationarity": *this is, again, a problem of scale.*

In order to grasp this idea in its utmost generality, everything needs to be molded in a complex form, because from the point of view of the theory of ensembles, which is the essential tool of interpretation, and to which a Madelung fluid makes direct reference, *the real numbers are not enough.* As Nicholas Georgescu-Roegen puts it with the occasion of an appropriate consideration of the problem of the theory of ensembles in connection with the theory of measure:

> Be this as it may, the actual difficulties — as I hope to show in this appendix — stem from two sources. Surprising though this may seem, the first source is the confusion created by the *underhand introduction of the idea of measure* at a premature stage in arithmetical analysis. The second source (an objective one) is the *impossibility of constructing a satisfactory scale of infinitely large and infinitely small with the aid of real numbers alone.* [(Georgescu-Roegen, 1971); *our Italics*]

Accordingly, Georgescu-Roegen offers an idea of construction of the m easure based on the concept of scale, which is quite natural, and even necessary we should say, in physics:

> ... To put it more generally, there are — as we shall argue presently — an infinite (in both directions) succession of classes of Numbers, *each class having its own scale*; every class is *finite* (original Italics!) *with respect to its scale, transfinite* (original Italics!) *with respect to that of immediate preceding class*, and *infrafinite* (original Italics!) *with respect to that of its successor.* Which scale we may choose as the finite scale *is a matter as arbitrary as the choice of the origin* of the coordinates on a homogeneous straight line. [(Georgescu-Roegen, 1971); *our Italics, except as indicated*]

It seems that physics realized up to this historical moment at least two of the requirements of this program, traced only symbolically, so to speak, in the work of Georgescu-Roegen just cited. First comes the interscale connection between *infinitely large* and *infinitely small* by the idea of a general statistic, whose positive expression is nowadays the Ehrenfest's theorem. Secondly, we have a specific meaning of the *arbitrariness of origin of a reference frame*, by the necessity of coupling between an event — an element of spacetime — with a 'position where the gravitational field is applied', in the phrasing of de Broglie. The expression of this coupling is nowadays the existence of general relativity. However, a suitable explanatory case of Georgescu-Roegen's general terms can only be properly made out of the Kepler problem, as applied to the classical planetary model of the atom. In making this case we need to use Hertz's terms 'material particle' and 'material point', or some other concepts, if we prefer [like Feynman 'partons' for instance; see (Feynman, 1969)], having nevertheless the same meaning, along the following line.

First, each material particle from the matter of the electron has a *transfinite* number of orbits to which it belongs: the ensemble of orbits corresponding to all material particles from the matter of nucleus. Accordingly, at the time scale of observed Kepler motion any material particle from the matter of electron belongs 'simultaneously' to every possible orbit due to the matter of nucleus. Here the term 'simultaneously' is interpreted in the sense that any material particle of electron matter belongs to such an orbit for an *infrafinite* time interval. Thus, when, in his Corollary 3 of Proposition VII, Problem II from the first book of *Principia*, Newton makes reference to the ratio of forces acting upon a material particle toward any point, these are not just arbitrary forces. Indeed, they are forces directed differently in space, but acting upon a particle moving along *the same Kepler orbit*, forasmuch as the requirement about the orbit is to be achieved 'in the same periodic times'. The Hertz type of natural philosophy liberates us from appealing to the idea of collisions, as Newton did, in order to physically justify the concept of force, or else to appeal to forces acting an infinitesimal time intervals, which are *infrafinite* in the sense of Georgescu-Roegen (Mazilu & Porumbreanu, 2018).

It is in this sense that a Hertz material particle can be characterized exactly in the manner in which Erwin Schrödinger describes the color continuum. We can just rephrase the words of Schrdinger, from the excerpt above, by replacing the appropriate words:

> ... the manifold of *material particles* has one of the powers of continuum, namely that of a space of functions, but the manifold of *material particle qualities* has the highest dimension three.

We can even name a principle analogous to that of "appearance on adjacent fields" here: it is the action of Newtonian type of forces having a magnitude inversely proportional to the square of distance between material particles. These are the gravitational forces, the electric forces and the magnetic forces. They act indeed selectively, from statical point of view, which means at any scale of space and time, otherwise they could not even be discovered as such. This further means that the material particles "arrange themselves into large groups, each one of the power of a function space". These are the Hertz's *material points*, forasmuch as, reproducing his own words:

> A material point ... consists of *any number* of material particles *connected with each other*. This number is always *infinitely great*: this we attain by supposing the material particles to be of *higher order infinitesimals than those material points which are regarded as being of infinitely small mass*. The masses of material points, and in especial the *masses of infinitely small material points*, may therefore bear to one another any rational or irrational ratio. [(Hertz, 2003), p. 46; *our Italics*]

Still paraphrasing Schrödinger, the manifold of these material points is... of the dimension three, and the problem is to construct such a manifold, a task that seems particularly easy for the selective forces we have in mind as 'connecting the particles with each other' in a material point.

A Characterization of the Hertz's Material Point

Fact is that in a static world, i.e. in a *world existing as such at any scale*, the three forces of Newtonian type must be in equilibrium. This observation should be able to allow us a connection between the mathematical order categorized by the sequence infrafinite — finite — transfinite, and the physical order categorized by the sequence microcosmos — macrocosmos — universe.

The main point of the theory of Berry and Klein on expanding force fields was to prove that the adiabaticity condition should not be a mandatory condition related to the idea of arbitrary phase in the wave mechanics (Berry & Klein, 1984). Incidentally, as we already noted before, this work touches a delicate issue related to cosmology: the length scale transformation — to use the terminology of Georgescu-Roegen in this specific instance — does not change the Newtonian forces of magnitude proportional to the inverse square of distance between classical material points. In other words, these forces are for the Newtonian cosmology just as significant as the blackbody radiation is for an Einsteinian cosmology. Indeed, we are now entitled to say, by virtue of Berry's and Klein's results, that an evolution of the universe, involving a transition of scale, leaves the Newtonian forces unchanged. This statement may seem particular because of the term 'Newtonian' to which it is referring. In hindsight though, we think that it is just as general as the Wien's displacement law in the case of blackbody radiation.

Indeed, until a proper analysis of the work of Berry and Klein, we can only notice now that it was inspired by the influence of expansion upon collisions. The forces involved in a collision, act between bodies only within infrafinite lengths of time, and therefore the theory of Berry and Klein, which carries over into a continuum the properties of collision forces, carries also the specific feature of a theory of transition between infrafinite and finite. Insofar as the Newtonian forces are physically the basic feature of the macrocosmos, therefore of a finite world, one can say that, as a mathematical tool, they reflect a finite mathematical order. Likewise, the blackbody radiation should be physically the mark of the universe at large, but mathematically it should be referring to a finite scale of things. This should be taken as the essential difference between the cosmology based on Newtonian forces and the cosmology based on blackbody radiation: the first one concerns the transition between infrafinite and finite, while the last one concerns the transition between finite and transfinite. From theoretical point of view, the key of physics here is the construction of a material point, which according to Hertz, contains "any number of material particles" provided this number is "infinitely great", in a space which is "infinitely small".

This last qualification sets a differentia of the concept of material point, by emphasizing its essential feature: *from physical point of view it is a transition concept.*

Classically, the image of a general material point is a finite solid body: its particles are at rest with respect to one another, thus making a fixed structure. We cannot say that this is the general idea of defining a material point, but only that it provides a limiting case, first because a body as presented to our senses is far from being an 'infinitely small space'. However, the Kepler problem as described above seems to indicate that those electrons of Madelung, which 'penetrate each other without merging', seem to qualify properly as such material points. First, every material particle entering the structure of an electron in revolution must be instantaneously, i.e. *at an infrafinite time scale with respect to the period of motion*, at rest. Such a situation has been aptly described by Joseph Larmor at the beginning of the last century:

> Imagine a *cloud of meteors pursuing an orbit in space*, under outside attraction — in fact, *in any conservative field of force*. Let us consider a group of the meteors around a given central one. *As they keep together, their velocities are nearly the same.* When the central meteor has passed into another part of the orbit, the surrounding region containing these same meteors will have altered in shape; it will in fact usually have become much elongated. If we merely *count large and small meteors alike*, we can define the density of their distribution in space, in the neighborhood of this group: *it will be inversely as the volume occupied by them*. Now, consider their *deviations from a mean velocity*, say of the central meteor of the group; we can draw from an origin a vector representing the velocity of each meteor, and the ends of these vectors *will mark out a region in the velocity diagram* whose shape and volume will represent the character and range of the deviation. [(Larmor, 1900); *our Italics*]

In order to set things in a right order here we just have to replace in Larmor's description the word 'meteor' with 'material particle', and thus describe an electron in revolution 'as a cloud of material particles pursuing a Kepler orbit in space'. Each material particle of the electron resides on a Kepler orbit characterized by a specific velocity, namely the velocity playing the part of initial condition as in equations (4.8−10). There will be, naturally, also a transfinite number of material particles residing instantaneously along the portion of that orbit located inside the electron. These are the material particles whose own orbits intersect the orbit of the chosen material particle. The term 'instantaneously' means here 'for an infrafinite time interval'. Thus, instantaneously, these particles are at rest with respect to each other. The situation can be easily described in the very same way in the whole space occupied by such an electron, whence the conclusion that the concept of material point asks for a 'static ensemble of material particles'. Insofar as the material particles have to be connected, according to the Hertz's definition of a material point, they need to be *connected by forces in equilibrium*. And as in the finite space range only the Newtonian forces are allowed in the Berry-Klein scaling, *these should be the forces responsible for the equilibrium inside a material point.*

Therefore, at this stage of the theory — the static stage, so to speak — we cannot use but the simplest of all dynamical principles — the Wigner's principle, forasmuch as each material particle needs to be at rest with respect to any other material particles. As each particle in this static stage is under the influence of all of the particles in the universe, regardless of scale, and as these forces should be in equilibrium, the Wigner's principle is to be used in the form: *the forces between material particles in matter always determine the velocities*, not the accelerations; only the forces acting upon particles inside component material points of matter that determine the accelerations, therefore the inertia.

Thus, the general plan of construction of the theory starts from the observation that a certain physical quantity is transported only by a mass transport: any other physical quantity is carried in a swarm of material points only together with their mass. Without mass there is no transport of anything else. As we shall see, the transport theory duly notices this important fact by a theorem defining the time of transport with the benefit of Lie derivative along the field of velocities of mass. Thus the mass itself, in a swarm of classical material points, can be concentrated two- dimensionally, like in the classical Kepler problem above, one-dimensionally, or zero- dimensionally, in well-defined mobile positions. Any other quantity though — such as the charge in both its instances, electric and magnetic, the color, the energy, the force, the momentum etc. — that has its own physical identity, is carried by the mass in a specific way to be described by the transport theory. In this context, a quantum jump does not necessarily mean a sudden jump in the value of the mass, as sometimes asserted: such a jump concerns *only the variation of a physical quantity carried by the mass current*, not the variation of the mass itself. In all instances, the Yang-Mills fields represent the appearance of this variation, which can be described by geometrical rules.

The General Meaning of Berry's Curvature

After all this discussion on the classical planetary model, we believe we are able to stage a scenario for a proper generalization of Berry's choice of the Hamiltonian from equation (6.8), with the assistance of a general presentation of the idea of spin. Take, first, the equation (6.9) without its initial connotation, i.e. as the intensity of a magnetic field: it is simply a vector field. Consider, as it was always the case in physics, that the slow coordinates are parameters that can be considered as functions of fast coordinates the the three-dimensional phase space. This, of course will imply some physical definitions of the slow coordinates as functions of the position in the phase space, but let us not get into details at this moment. The projection of the Liouville volume (6.10) along the field (6.9), which is instrumental in the definition of the time derivative in this formalism, is given by a 2-form:

$$i_B(\Omega) = g \frac{R_1 dx^2 \wedge dx^3 + R_2 dx^3 \wedge dx^1 + R_3 dx^1 \wedge dx^2}{R}. \tag{6.40}$$

This differential form fits into Dumachev's scheme (6.15-17) only if the 1-form (6.16) is built upon a vector $\mathbf{h}$, solution of the vectorial equation

$$\nabla_R \times \mathbf{h} = g \frac{\mathbf{R}}{R^3}. \tag{6.41}$$

Consequently, only along the solutions of this differential equation is the elementary Liouville volume preserved. We can give a functional estimate of this field, for it is directed radially with respect to the origin of positions in the slow parameters' space. So, in appropriate spherical coordinates we have

$$(\nabla_R \times \mathbf{h})_R = \frac{g}{R^2}; \quad (\nabla_R \times \mathbf{h})_\theta = 0; \quad (\nabla_R \times \mathbf{h})_\varphi = 0. \tag{6.42}$$

If the radial component of the vector $\mathbf{h}$ is constrained to be an arbitrary function of the radial coordinate alone, the two last equations here become

$$R \frac{\partial h_\theta}{\partial R} + h_\theta = 0, \quad R \frac{\partial h_\varphi}{\partial R} + h_\varphi = 0. \tag{6.43}$$

This means that both components of the vector field $\mathbf{h}$ involved in these differential equations are homogeneous functions in R of degree -1:

$$h_\theta = \frac{A(\theta, \varphi)}{R}, \quad h_\varphi = \frac{B(\theta, \varphi)}{R} \tag{6.44}$$

with A and B arbitrary functions of their arguments. Thus, the first equation (6.42) reduces to

$$\frac{\partial (B \sin \theta)}{\partial \theta} - \frac{\partial A}{\partial \varphi} = g \sin \theta. \tag{6.45}$$

Now, assume a further simplification, looking for the field having $A(\theta, \phi)$ as a function of θ only, $H(\theta)$ say. Then we can solve for $B(\theta, \phi)$, with the general result

$$B(\theta, \varphi) = -g \cot \theta + \frac{G(\varphi)}{\sin \theta} \tag{6.46}$$

108

where $G(\phi)$ is an arbitrary function of its argument. Thus, we have a solution of (6.41) whose components in a spherical coordinate system are:

$$H_R = F(R); \quad h_\theta = \frac{H(\theta)}{R}; \quad h_\varphi = -\frac{g}{R}\cot\theta + \frac{G(\varphi)}{R\sin\theta} \quad (6.47)$$

with F, G, H arbitrary functions of their unique arguments. The differential form 'h' is not an exact differential, as expected of course:

$$h = h_R dR + h_\theta R d\theta + h_\varphi R\sin\theta d\varphi$$
$$= F(R)dR + H(\theta)d\theta + G(\varphi)d\varphi - g\cos\theta d\varphi. \quad (6.48)$$

However, we can characterize its departure from that condition by a known differential form:

$$-g\cos\theta d\varphi = -\frac{g}{R}z\frac{xdy - ydx}{x^2 + y^2}. \quad (6.49)$$

This differential form characterizes the Schwinger's description of a magnetic monopole [(Felsager, 1983), Chapter 9]. If F and H are zero and G is a constant, we have the *Dirac's monopole*. This situation was indeed, as it is actually even today, interpreted in terms of a magnetic monopole.

According to the formalism just presented, a special motion can be obtained, by further considering the 2-form (6.40) as a fundamental 'Liouville volume', as it should be in view of its geometrical meaning. Indeed, $i_B(\Omega)$ is simply proportional to the elementary oriented area in the fast variables, which suggests an elementary solid angle. Assuming variations in the slow parameters in order to set the scale of time variation as in quantum mechanics (Mazilu & Porumbreanu, 2018), the projection of the flux (6.40) along the velocity vector field $\dot{\mathbf{R}}$:

$$i_R[i_B(\Omega)] = g\frac{\mathbf{R} \times \dot{\mathbf{R}}}{R^3} \cdot d\mathbf{r} \quad (6.50)$$

gives a rate of the solid angle in the slow variables. This is the elementary work of the force, anticipated by Henri Poincaré in describing the *action of a magnetic pole upon an electric charge*, in order to explain the results of the experiments of Kristian Birkeland (Poincaré, 1896). However, as we shall shortly see here, it can be rightfully connected to a physical deformational process, both geometrically and physically.

Chapter 7. The Physical Point of View in the Theory of Surfaces

We start from the idea that a surface should play a double role in physics. First, it plays the obvious role of separator, specifically for separating the matter from space or of the space from matter, for these are altogether two different situations. This statement cannot be made understandable without the second role of the surface which is closer to the regular mathematical role: that of *locus of an ensemble of positions* in the geometrical acceptance. It pays, however, to notice that it is in this instance that the surface allows the mathematical description and the consequent physical understanding of that property of the material points of 'penetrating without merging', to use the phrasing of Erwin Madelung. In order to get a better grip of this idea, some introductory observations seem to be in order here.

A Few Mathematical Prerequisites

It seems to us obvious that a physical theory of surfaces cannot be developed but by the idea of embedding. As a matter of fact, the embedding is an interpretation and, forcing a little the idea, the essential point of the Darwin's definition of interpretation is the existence of an ensemble of positions giving the possibility of interpretation of the surface as a geometrical locus. Here a position can be taken geometrically as a point and physically as a material particle in the sense of Hertz. In this last instance, the material particle is to us a little more than it was to Hertz himself, in the sense that we added to it the feature of *indicator of a point in space*, by a process of application of a field to a position, being it in space or in matter in a sense which will be explained here shortly. For instance, in keeping with the classical physics, one can imagine the point of application of a force or of a resultant of many forces, according to the third principle of classical dynamics, but this is quite a particular example.

The mathematical method itself, for carying out the task of embedding a surface, and therefore the task of interpretation in general, is based on an almost trivial statement that emerged apparently largely unnoticed, or if noticed, it has not been properly used for physical purposes. In order to make it obvious, we reproduce here two of the Élie Cartan's algebraical theorems which form the ground of his approach of differential geometry by *moving frames* [for a clear description of the idea from the point of view we adopt here, see (Spivak 1999), Volume II, Chapter 7]. These theorems are borrowed directly from one of Cartan's courses, published via the Russian geometrical school of S. P. Finikov [(Cartan, 2001); pp. 16 17, Theorems 7 & 9]. We just appropriate them here under the name of *Cartan Lemmas 1 and 2*, only in order to be suitably used in making our point as explicit as possible:

Lemma 1. Suppose that $s^1, s^2, ..., s^p$ is a set of linearly independent 1-forms. Then we have

$$s^\alpha \wedge \phi_\alpha = 0 \quad \leftrightarrow \quad \phi_\alpha = a_{\alpha\beta} s^\beta \quad , \quad a_{\alpha\beta} = a_{\beta\alpha}$$

where $\phi_1, \phi_2, ... \phi_p$ is another set of linearly independent 1-forms, and summation over repeated indices is understood.

Lemma 2. Suppose the basis differential elements $du^1, du^2, ..., du^n$ are connected by a system of equations

$$\omega^1 = 0, \quad \omega^2 = 0, ..., \quad \omega^p = 0$$

where ω^α, $\alpha = 1, 2, ..., p$ are linearly independent 1-forms. Then the 2-form 'f' *vanishes as a consequence of this system* if, and only if, 'f' can be written as

$$f = \omega^\alpha \wedge \phi_\alpha$$

where, again, summation over α is understood, and ϕ_α are p *conveniently chosen* 1-forms.

The first one of these theorems is, by and large, known as *Cartan's Lemma* proper in the specialty literature, while the second one carries no specific name. What seems to be essential in these lemmas, and is always stressed in many old treatises, but apparently forgotten lately — perhaps due to exclusive mathematical applications — is the fact that the symmetric matrix **a** from Lemma 1, as well as the 1-forms ϕ_α from Lemma 2, are *things external to the geometrical problem*, things through which the physics can therefore be introduced. In concentrating on the local description at a point of a surface, without being interested of the global aspects of that surface, as it is almost always the case in physics, this observation is essential. In this respect another convention referring to our use of indices is again in order: insofar as the space contemplated as environment in the embedding problem necessary to physical interpretation is here the usual, possibly Euclidean, three-dimensional space of our intuition, we reserve the Latin indices exclusively for this case. The Greek indices are used for any other dimension, as in the case above, but especially for dimension two, in the case of surfaces, and four in the case of the manifold of events — the spacetime.

Now, from the point of view of physics, these two theorems of Cartan are not sufficient. Take for instance the extended electron model of Dirac, which can be considered as the epitome of a general model of *extended material particle* (Dirac, 1962), a point of view advocated in fact relatively long ago by Malcolm Mac Gregor (Mac Gregor, 1992). This is a matter sphere embedded in space, which needs to be characterized from electromagnetic point of view. And in the classical electrodynamics the relation between matter and space is resumed by electromagnetic constitutive laws, which primarily involve two exterior differential 2-forms and two exterior 1-forms (van Dantzig, 1934). In this case, a generalization of Cartan's Lemma 1 is necessary, which must be able to allow us distinguish the presence of fields in matter and their correlation with the fields in space. This generalization has already been around for a relatively long time in the mathematical specialty literature, under the name of Yoshio Agaoka, through the following theorem (Agaoka, 1989):

Assume that π_α are 'r' exterior differential 1-forms representing a coframe with respect to a r- dimensional manifold in space. If 'r' differential p-forms ω^α satisfy to equation

$$\omega^\alpha \wedge \pi_\alpha = 0$$

then the (p-1)-forms $\gamma^{\alpha\beta}$ exist such that

$$\omega^\beta = \gamma^{\beta\alpha} \wedge \pi_\alpha; \quad \gamma^{\beta\alpha} = \gamma^{\alpha\beta}$$

for $\alpha, \beta = 1, 2,..., r$.

The Cartan's lemma 1 can be derived from this theorem of Agaoka for p = 1, in which case $\gamma^{\alpha\beta}$ must be 0-forms, i.e. simple functions. The case p = 2 is 'maximal', so to speak, in three dimensions, for in that case there are no exterior differential forms of degree higher than three. Geometrically, the best candidate for the matrix γ seems to be a 2×2 matrix representing the variation of the curvature matrix or, more generally, a matrix related to it. Postponing a little bit the description of general theory of surfaces from this point of view, an account of the physical example based on the Dirac's model of space extended electron (Dirac, 1962) seems in order. This is intended to clarify the position of Agaoka's theorem *per se*, outside the framework of the Cartan's lemmas. Our rendering here does not follow the details of the original description. We only build upon the idea of geometrical shape of matter in a *space extended form*, highlighting the needs of a classical electromagnetic description, just to make a point.

Assume that the limit separating the matter from space is indeed a surface, generalized in the following way: it is a proper geometrical surface, but from physical point of view it is also a support of 'dual physical properties',

112

i.e. properties representative to matter as well as properties representative to space. In other words, here, on this surface, the matter meets the space, and physics needs to describe this 'collision of the two worlds', so to speak. In an electromagnetic structure, the space is simply represented by a 2-form, 'b' say, while the matter is represented by a 2-form, 'd' say. These symbols suggest electromagnetism in the sense of David van Dantzig, as representative for the purely electromagnetic structure of matter (van Dantzig, 1934). To wit, these exterior differential 2-forms can indeed be appropriated here as two forms denoted above ω^{ν} for a general formulation of theory:

$$\omega^1 \equiv b = b_{ij}dx^i \wedge dx^j; \quad \omega^2 \equiv d = d_{ij}dx^i \wedge dx^j.$$

They correspond to *magnetic induction* and respectively *electric induction*, from the classical theory of electricity and associated magnetism. An explanation is in order: if the surface delimiting matter can be globally defined, as in the case of Dirac's electron for instance, these two exterior forms satisfy the integral equations:

$$\oiint_{Surface} b_{ij}dx^i \wedge dx^j = 0;$$

$$\oiint_{Surface} d_{ij}dx^i \wedge dx^j = \iiint_{Volume} \rho_{ijk} dV^{ijk}$$

which justify their names, insofar as the magnetic field is source-free while the electric field has the electirc charge as its source. Here ρ is the third order tensor of the *electric density of matter*, while $d\mathbf{V}$ is the oriented volume 3-form in the space occupied by matter. These relations can be used to justify our choices for describing the matter and space in their common limit.

Indeed, in spite of the global aspect of the problem, as the previous integral equations seem to suggest, the physical definition of the two fields involved here *is always local*, at least experimentally. To wit, we have the infinitesimal *voids* in the magnetic matter for defining 'b', or the *dipoles* in the electric matter for defining 'd'. This means that the tensor **b** should be always realizable *in matter* as a vector, say **h**, defining the 1-form of the magnetic field. Likewise, the tensor **d** is representable *in space* by a vector, **e** say, defining the 1-form of electric field. It is here the point where a theorem which we termed as *Agaoka's theorem*, makes its mark of necessity, because we need a *constitutive law* relating the fields **b** and **d** to the fields **e** and **h**. Thus, between the corresponding differential forms we always need a general relationship of the form:

$$b \wedge h + d \wedge e = 0$$

for only in this case, *according to Agaoka's theorem*, there should be a matrix of exterior differential 1-forms, γ say, which realizes a linear correlation between

113

the 2-forms and 1-forms:

$$b = \gamma^{11} \wedge h + \gamma^{12} \wedge e; \quad d = \gamma^{12} \wedge h + \gamma^{22} \wedge e.$$

In terms of the field vectors themselves, these equations can be turned into linear constitutive equations of the kind used in the classical theory of electromagnetism. The exhaustive characterization of electromagnetism from this point of view is due to Reginald Aubrey Fessenden, at the beginning of the last century (Fessenden, 1900). But the situation is quite general, so that now, after this introduction of the mathematical basis and the reasons for such an approach to a physics of surfaces, we need to present in detail the essence of the geometric theory necessary to achieve this physics.

The Differential Theory of Surfaces

Élie Cartan's exterior differential approach to the theory of surfaces, allows us to say that the local differential theory of surfaces is simply a consequence of the fact that the elementary (differential) displacement of a point of surface is an intrinsic vector of that surface (Guggenheimer, 1977). Everything in the differential theory of surfaces, and in the embedding theory in general, follows from this simple fact through the rules of handling of the differential forms. To wit, if $\mathbf{x}$ is the position vector of a point of surface, then $d\mathbf{x}$ should be an intrinsic vector of the surface, at least as long as no phenomenon occurs forcing us to consider that the position leaves the surface. But even in such a case, important from a physical point of view, the elementary variation of position *in space* can be described by a process of deformation through a continuous family of surfaces containing the position during that variation. At first, the deformation appears as a purely mathematical device, but as soon as we get into details it becomes gradually obvious that it needs further consideration, especially from physical point of view.

Let us now lay down the proper mathematical form of this portrayal of surfaces. Referring the local geometry at a position $\mathbf{x}$ in space to an orthonormal frame of reference: $(\hat{\mathbf{e}}_1, \hat{\mathbf{e}}_2, \hat{\mathbf{e}}_3) \equiv \langle \hat{\mathbf{e}} |$, we write

$$d\mathbf{x} = s^1 \hat{\mathbf{e}}_1 + s^2 \hat{\mathbf{e}}_2 + s^3 \hat{\mathbf{e}}_3 \quad \leftrightarrow \quad d\mathbf{x} = \langle s | \hat{\mathbf{e}} \rangle. \tag{7.1}$$

These relations call for a further elucidation regarding the notations. A caret over a bold symbol means a unit vector. Further on, a 'Dirac representation' with 'kets' and 'bras' is used for any sequence of mathematically meaningful objects. For instance, $|s\rangle$ here represents the matrix 3×1 having the components of $\mathbf{x}$ as entries. It is thus understood that the vector notation $\mathbf{x}$ involves also the reference frame $|\hat{\mathbf{e}}\rangle$ in its definition, which, according to our notation is a 3×1 matrix having the unit vectors of reference frame as components. The last equality in equation (7.1) embodies such a complete definition for the elementary geometrical displacement. A certain vector can definitely be represented just as a 'ket' or a 'bra', either in the same reference frame, or globally, in cases where the reference frame does not matter, as in the case of an Euclidean universe.

From the differential geometric point of view, any portion of a surface defined by some kind of functional continuity is locally describable by means of two elementary differentials s^1 and s^2, and a direction defined by a unit vector, normal to surface element. This unit normal is bound to retain its normalization over the entire surface element, so that it the normalization condition can be taken as a defining property of the very surface element. In this case, the surface element can be parametrized in a manner that opens the gateway for physics. First, let us refer the surface to the ambient space containing the element of surface, and described geometrically in some general coordinates $(x^k, k = 1, 2, 3)$ gauged by means of a certain Euclidean reference frame. The elementary displacement from the second equation of (7.1), will be written again in the form involving the space coordinates, in order to make the idea of embedding more explicit:

$$d\mathbf{x} = (dx^k)\hat{\mathbf{e}}_k \tag{7.2}$$

where the summation convention over repeated indices of different variance is used. This vector should be an exact differential, so we must have: $d \wedge d\mathbf{x} = \mathbf{0}$, which means

$$(d \wedge dx^k)\hat{\mathbf{e}}_k + d\hat{\mathbf{k}} \wedge dx^k = \mathbf{0}. \tag{7.3}$$

Now, assuming a Frenet-Serret evolution of the reference frame, we have

$$d\hat{\mathbf{e}}_k = \Omega_k^j \hat{\mathbf{e}}_j; \quad \Omega_k^j + \Omega_j^k = 0, \tag{7.4}$$

the last condition here being a consequence of the orthonormality of frame. Putting this into (7.3), gives the compatibility equations

$$\langle d \wedge dx| + \langle dx| \wedge \mathbf{\Omega} = \langle 0|. \tag{7.5}$$

Perhaps it is the case to insist again just a little bit upon our choice of symbols here. Namely, the symbol '$d\wedge$' means *exterior differentiation* in order to distinguish this type of differentiation from the regular differentiation. The simple sign '$\wedge$', taken by itself, means only exterior product of the differential forms.

Now, assume that this reference frame of ambient space is adapted to a surface, in the sense that we can describe the space from the neighborhood of any point of surface by taking $\hat{\mathbf{e}}_3$ as the unit normal of surface in that point, and $\hat{\mathbf{e}}_1, \hat{\mathbf{e}}_2$ as unit vectors on the surface. This, of course, entails the identities

$s^k \equiv dx^k$ as in (7.1), in which case *the surface itself can be described as embedded in the ambient space by the condition $s^3 = 0$*. This condition does not mean too much by itself, but when referred to equation (7.5) above, it has a few important consequences. First, the Frenet-Serret equations (7.4) can be written as:

$$d\hat{\mathbf{e}}_1 = \Omega_1^2 \hat{\mathbf{e}}_2 + \Omega_1^3 \hat{\mathbf{e}}_3, \quad d\hat{\mathbf{e}}_2 = \Omega_2^1 \hat{\mathbf{e}}_2 + \Omega_2^3 \hat{\mathbf{e}}_3, \quad d\hat{\mathbf{e}}_3 = \Omega_3^1 \hat{\mathbf{e}}_1 + \Omega_3^2 \hat{\mathbf{e}}_2 \quad (7.6)$$

situated on surface have also components along the normal to surface, obviously due to the curvature of the surface. This is why the entries Ω_1^3 and Ω_2^3 of the matrix $\boldsymbol{\Omega}$ are usually taken *as measures of the curvature of surface*. In the case considered here they can be put naturally in connection with external, perhaps physical conditions, which might be able to locally determine the curvature of surface, or at least only its variation. These conditions come out from equation (7.5), which, combined with the embedding condition $s^3 = 0$, gives

$$d \wedge s^1 + s^2 \wedge \Omega_2^1 = 0, \quad d \wedge s^2 + s^1 \wedge \Omega_1^2 = 0, \quad s^1 \wedge \Omega_1^3 + s^2 \wedge \Omega_2^3 = 0 \quad (7.7)$$

or in index notation with the summation convention:

$$d \wedge s^\mu + s^\nu \wedge \Omega_\nu^\mu = 0, \quad s^\beta \wedge \Omega_\beta^3 = 0. \quad (7.8)$$

This equation is referring exclusively to the surface, viz. to its intrinsic geometry. This means that the indices take just the values 1 and 2, which explains the Greek symbols for indices.

The last equation (7.7) shows that between the components of the vector $|\Omega^3\rangle$ representing the curvature of surface, and the vector $|s\rangle$ of elementary displacements on the surface, there is a linear homogeneous relation. This is guaranteed by the Cartan's Lemma 1, which says that the last equation (7.7) is equivalent to existence of a convenient symmetric matrix, $\mathbf{b}$ say, determining the components of the curvature vector with respect to the variation of position by a linear homogeneous relation:

$$|\Omega^3\rangle = \mathbf{b} \cdot |s\rangle \quad \therefore \quad \Omega_\nu^3 = b_{\nu\sigma} s^\sigma. \quad (7.9)$$

The equations above suffice in making a point about the place of physics in determining the local geometry of a surface. It is offered by the entries of the symmetric matrix $\mathbf{b}$, which can be properly called a *curvature matrix*: these can be *any three convenient numerical things* – the curvature parameters – provided they are arranged in a 2×2 symmetric matrix. The trace of the curvature matrix gives the *mean curvature*, and its determinant gives the *Gaussian curvature*. Our contention here, inspired by this way of introducing the curvature parameters, is that even these measured quantities are actually *external parameters*, possibly physical, a feature bestowed upon them by the Cartan's

Lemma 1. Fact is that they are always subjected to variations when physical causes, like the contact of bodies for instance, participate in changing the local aspect of their surface. However, the evaluation of the curvatures − in case one needs it, and has the conditions to do it, of course − can be done just by geometrical measurements on the surface, as in the usual classical procedure (Lowe, 1980).

In this respect, another point is worth making, for the very sake of physics: if we have at our disposal a meaningful unit vector, or simply a meaningful vector for that matter, as in the case of Louis de Broglie's theory of waves associated with the physical rays (de Broglie, 1926), we can always connect to this vector an element of surface by the condition of orthogonality. Indeed, let $\hat{\mathbf{u}}$ be a unit vector with origin in the space position $\mathbf{x}$. Then the surface whose normal is $\hat{\mathbf{u}}$, can be defined by the condition that $d\mathbf{x}$ should be normal to $\hat{\mathbf{u}}$, while $d\hat{\mathbf{u}}$ should be an intrinsic vector:

$$d\mathbf{x} \cdot \hat{\mathbf{u}} = 0, \quad d\hat{\mathbf{u}} = -\mathbf{a} \cdot |s\rangle \qquad (7.10)$$

where $\langle s| \equiv (s^1, s^2)$ is the 'bra' having the surface displacements as entries, and $\mathbf{a}$ is a 2×2 matrix. Of course, this matrix should satisfy some constraints, based on the relation between $d\mathbf{x}$ and $|s\rangle$. However, insofar as only the relations from equation (7.10) are involved, all we can say is that only the symmetric part of the matrix $\mathbf{a}$ can be taken as representing the curvature matrix of the surface. Indeed, we have

$$d(d\mathbf{x} \cdot \hat{\mathbf{u}}) = d^2\mathbf{x} \cdot \hat{\mathbf{u}} + d\mathbf{x} \cdot d\hat{\mathbf{u}} = 0 \quad \therefore \quad d^2\mathbf{x} \cdot \hat{\mathbf{u}} = \langle s|\mathbf{a}|s\rangle \qquad (7.11)$$

which proves our statement, and the choice of minus sign in (7.10) in order to give an adequate quadratic form in (7.11). This last equation shows that, as long as only physics is concerned, the curvature matrix *may not be necessarily symmetrical*, because it is not decided by Cartan's Lemma 1, but by the linear dependence defining the surface element with respect to a given vector in space. This is, for instance, the case of mechanical torsion of plates (Lowe, 1980), and thereby we can guess that such a situation involves forces, an idea that we shall subsequently develop here.

The equations above suffice in order to settle the ideas with respect to a physical perception of space, which seems to be liable to be made into a general physical theory. The man did not gain the concept of space out of nothing: he needed a nest to grow in, and this nest is provided by the Earth. The idea of space has grown along the time starting from its embryo − the space of our intuition. This space occurred in turn only because it has been made possible by the fact that the man grows on a quintessential surface of separation: the surface of Earth. In the interest of physics, it becomes therefore necessary to show how a space can be described with respect to a given surface, just as we have shown how the surface can be with respect to a given space.

Rainich's Description of Surrounding Space

The equation (7.10) may appear as instrumental to a physical theory aiming to locally constructing a surface. This means that the equations of embeding the space $s^k \equiv dx^k$ may not be satisfied, and thus the differentials on surface are, in a sense, arbitrary with respect to space coordinates: they need to be chosen in advance and, thus, the surrounding space to be described with respect to surface by a reference frame that needs to be constructed *ad hoc*, not simply *adapted* from among those of surrounding space. The situation appears at its best if we analyze the condition of embedding used thus far. Indeed, the embedding condition $s^3 = 0$, used to obtain the last equation in (7.7) or (7.8) is quite particular from the point of view of the differential calculus. Indeed, those equations represent actually the condition $d \wedge s^3 = 0$, which is satisfied even for s^3 *constant but arbitrary*, i.e. for any surface parallel to the one given by $s^3 = 0$. In fact, the most general solution of the equation $d \wedge s^3 = 0$, is *an arbitrary exact differential* 1-form. In view of this, further differential conditions should be provided in order to describe the embedding a surface in the surrounding space. In the classical theory of surfaces these conditions are usually known as Gauss-Codazzi, or sometimes as Mainardi-Codazzi equations. We will illustrate their necessity by the way of an example involving explicitly the physics from the very beginning of the construction. Obviously, this example mimics appropriately the very historical order of ways in which the man gained to concept of space.

Alois Švec insisted upon the fact that an adaptation of a geometrical theory of space to a given surface is fundametal (Švec, 1988), but in such a process essential would not be so much the adaptation to surface of a reference frame, as much as a kind of *adjustment to the parametrization of the surface*. The most general kind of adjustement according to Švec's idea, is linear, and therefore can satisfy a differential equation of the type:

$$s^1 \wedge du + s^2 \wedge dv = 0. \qquad (7.12)$$

According to Cartan Lemma 1, the fundamental differentials s^1 and s^2 are then linear in the differentials of the functions 'u' and 'v' − thc parameters on surface. Indeed, according to Lemma 1, from equation (7.11) we have that the elementary displacements must be expressed in terms of the differentials (du) and (dv) by

$$s^1 = \alpha du + \beta dv; \quad s^2 = \beta du + \gamma dv. \qquad (7.13)$$

118

The original case of Alois Švec (loc. cit.) is included here for a special choice of the parameters, whereby $\beta = 0$. Thus, the fundamental differentials on our surface — that, in this instance, can be properly called *surface of reference* — which are not always exact differentials, are nevertheless always linear combinations of such differentials.

Now, from (7.13) we get by exterior differentiation:

$$d \wedge s^1 = \frac{\gamma d\alpha - \beta d\beta}{\alpha\gamma - \beta^2} \wedge s^1 + \frac{\alpha d\beta - \beta d\alpha}{\alpha\gamma - \beta^2} \wedge s^2;$$

$$d \wedge s^2 = \frac{\gamma d\beta - \beta d\gamma}{\alpha\gamma - \beta^2} \wedge s^1 + \frac{\alpha d\gamma - \beta d\beta}{\alpha\gamma - \beta^2} \wedge s^2,$$

(7.14)

so that a system corresponding to (7.8) can be obtained from this by the identification

$$\mathbf{\Omega} \equiv (\mathbf{d}\xi)\mathbf{1} + \begin{pmatrix} -\omega^2/2 & \omega^1 \\ -\omega^3 & \omega^2/2 \end{pmatrix}$$

(7.15)

where the three differential forms $(\omega^1, \omega^2, \omega^3)$ represent a coframe of the **sl** (2,R) algebra:

$$\omega^1 = \frac{\alpha d\beta - \beta d\alpha}{\Delta}; \quad \omega^2 = \frac{\alpha d\gamma - \gamma d\alpha}{\Delta}; \quad \omega^3 = \frac{\beta d\gamma - \gamma d\beta}{\Delta};$$

$$\Delta = \alpha\gamma - b^2$$

(7.16)

and we denotd $\xi \equiv \ln\sqrt{(\Delta)}$.

Represented as such, *in the manner of Švec* we should say, the geometry of space in the neighborhood of a position from the surface taken as reference, is a conform euclidean geometry, completely determined by the surface and its curvature matrix at the given position. Indeed, the position of a point in surrounding space can be located in a local Euclidean reference frame by the differential coordinates (Rainich, 1925)

$$dx = \frac{1}{2}[(1 - u^2 + v^2)s^1 - 2uvs^2],$$

$$dy = \frac{1}{2}[-2uvs^1 + (1 + u^2 - v^2)s^2],$$

(7.17)

$$dz = us^1 + vs^2.$$

The parameters 'u' and 'v' are convenient parameters on the surface. The origin of reference frame is physically defined by the equations $s^1 = s^2 = 0$. Thus, the qualification 'physically' means here that from a mathematical point of view, in this definition there should be some relations between the parameters (u, v) and the parameters (α, β, γ), whose parentage can be physical. Further on,

(dx) and (dy) from equation (7.17) represent homogeneous transforms of the differential forms $s^{1,2}$, with the help of the surface parameters themselves.

This suggests a general philosophy related to this representation: the surface 'arranges' locally the physics — as it always did, in the historical order mentioned above — in the sense that the accomodation of space to surface is first given by the condition (7.13) on differentials, which need themselves to express the surface physics in a certain way. This very physics is then lent to space by a condition of accomodation of the differentials of the coframe:

$$dx \wedge s^1 + dy \wedge s^2 = 0. \tag{7.18}$$

According to Cartan's Lemma 1, this equation leads to $|dx\rangle = \mathbf{m} \cdot |s\rangle$, where $\mathbf{m}$ is here a symmetric 2×2 matrix having *entries which depend exclusively on surface*. These entries are concretely expressed by some quadratic forms in the binary parameter (u,v), as shown in the first two equations from (7.17). On the other hand, (dz) clearly represents the height of the point in space with respect to the tangent plane of the surface, given by the second fundamental form of the surface in case the matrix $\boldsymbol{\alpha}$ determining the transformation (7.13) is the curvature matrix of the surface. Obviously this is the case with equation (7.18). Indeed, the first fundamental form of the surface is given by the Euclidean metric of the surrounding space, which can be obtained directly from equation (7.17). The metric tensor turns out to be:

$$\lambda^2 \begin{pmatrix} \alpha^2 + \beta^2 & \beta(\alpha + \gamma) \\ \beta(\alpha + \gamma) & \beta^2 + \gamma^2 \end{pmatrix} \equiv (\lambda\boldsymbol{\alpha})^2 \tag{7.19}$$

with $2\lambda \equiv 1 + u^2 + v^2$. Classically speaking, this tensor represents actually *the third fundamental form* of a regular surface, just as (dz) is its *second fundamental form*. The conditions that the differential forms (7.13) are exact differential can be written in the form:

$$d \wedge s^\alpha = \Theta_\gamma^\alpha \wedge s^\gamma;$$
$$\Theta \equiv \frac{1}{1 - \mu^2 - \nu^2} \begin{pmatrix} d\mu + \mu d\mu + \nu d\nu & d\nu + \mu d\nu - \nu d\mu \\ d\nu - (\mu d\nu - \nu d\mu) & -d\mu + \mu d\mu + \nu d\nu \end{pmatrix} \tag{7.20}$$

with $\mu = u^2 - v^2$ and $\nu = 2uv$. When written in the parameters 'u' and 'v', they lead to the Gauss-Codazzi equations, which impose a geometric continuity on surface for the physical parameters (α, β, γ), characterized by the partial differential equations given by Rainich himself (*loc. cit.*):

$$(1 + u^2 + v^2)(\partial_v \alpha - \partial_u \beta) + 2(\alpha + \gamma)v = 0,$$
$$(1 + u^2 + v^2)(\partial_u \gamma - \partial_v \beta) + 2(\alpha + \gamma)u = 0, \tag{7.21}$$
$$u(\partial_v \alpha - \partial_u \beta) = v(\partial_u \gamma - \partial_v \beta).$$

One can, therefore, rightfully say that these equations accommodate the physical parameters to the given surface.

A Physical Parametrization of Surface

As we have seen, the surface on which a Kepler orbit lies can be fully described by deformations. As long as the physics is involved here, the deformation is generated by forces, and these forces can be taken as deciding the surface parametrization necessary to a 'Švec stage', so to speak, in defining the physics of surface. The forces we are talking about here should be valid in any initial conditions of the kind involved in expressing the trajectory in the classical Kepler problem. Put otherwise, the mathematical expression of these forces should be valid no matter of the material particle chosen from a 'swarm' representing, in the sense of Madelung, the orbiting matter to which they are applied according to the third principle of dynamics. This is a property of invariance which needs to be taken into consideration in the choice of the parameters on the reference surface itself. A specific choice is of particular interest in illustrating the manner in which that property of invariance should be understood.

The motions from which the initial conditions of a Kepler problem are chosen should be of the same nature as the actual motion which they help describe through the equations of the classical dynamics. This observation can be mathematically used for a conclusion involving forces, as long as these are central (Appell, 1889, 1891). Thus, if 'u' and 'v' are the plane coordinates of some classical motion under a central force, then their equations describing that motion are

$$\frac{d^2u}{d\tau^2} = F\frac{u}{\rho}; \quad \frac{d^2v}{d\tau^2} = F\frac{v}{\rho}; \quad \rho^2 = u^2 + v^2 \qquad (7.22)$$

where F is the magnitude of that force and τ is the time of motion. This is simply a convenient transliteration of a dynamical equation as, for instance, (4.1) in the case of Kepler problem, but for a general force, not necessarily a Newtonian one. It is intended to help represent the coordinates of the plane of motion under a central force as the *coordinates on a surface* having this plane as a tangent plane, or as a plane in some relation with the tangent plane, for instance parallel to it. Based on equations (7.22), if the force is central, we can construct the area integral expressing the second of the Kepler laws (4.3), which in this case is:

$$u\frac{dv}{dt} - v\frac{du}{dt} = \dot{a}.$$

Now, assume that these are the coordinates which, as shown beforehand, give a content to the time and radial coordinate in a gravitational field. Specifically, as suggested by the transformation (2.41), we take 'u' and 'v' as factors of the ratios involved in the group transformations, i.e.

$$r = \frac{u}{v}, \quad z = \frac{1}{v}. \tag{7.23}$$

Then, the equations of motion (7.22) become

$$\frac{d^2 r}{d\tau^2} = 0; \quad \frac{d^2 z}{d\tau^2} = -F\frac{v^3}{\rho}; \quad d\tau = \frac{dt}{v^2}. \tag{7.24}$$

Now, the radial coordinate from equation (2.41) is taken as the 'vertical' coordinate, which explains our notation by 'z'. On the other hand, the first equation (2.41) gives a 'horizontal' coordinate in an arbitrary direction, denoted here with 'r', so that the (r, z)-coordinates can be taken as some cylindrical coordinates. The equations (7.24) then suggest a free fall in a vertical force field, exactly as in the case of gravitational motion close to Earth surface, whereby the gravitational force can be practically considered as a field with parallel vertical lines of force. Appell's procedure is based upon the fact that if the point (u, v) describes a conic, the point (r, z) describes a conic too, the homographic transform of the first one. This observation is particularly important in the problem of finding the central forces corresponding to given orbits, which is thus reduced by equation (7.24) to the problem finding those 'parallel fields of forces' of magnitude $Z = -Fv^3/\rho$, whose point of application describes a conic. This problem can, in turn, be further reduced to that of solving the *Halphen's equation of conics* in the plane of coordinates (r, z) (Halphen, 1879).

Indeed, from equations (7.24) we have

$$\frac{d^2 z}{d\tau^2} = \frac{1}{\dot{a}^2} Z.$$

Now, it is clear that any differential equation for the coordinate 'z' as a function of 'r', become here an equation for the parallel force Z as a function of 'r'. In particular we have:

$$\{(z)^{-\frac{2}{3}}\}''' = 0 \quad \leftrightarrow \quad \{(Z)^{-\frac{2}{3}}\}''' = 0$$

where an accent means derivative with respect to 'r', and the first equation is Halphen's equation of a conic in the plane (r, z). Therefore the parallel force is

$$Z = \mu \phi^{-\frac{3}{2}} \quad \text{where} \quad \phi''' = 0$$

with μ — a constant. Paul Appell found this way two expressions of the parallel force, valid *in any initial conditions* for the motion:

$$Z = \frac{\mu c^{\frac{3}{2}}}{(br + cz + e)^3} \quad \text{and} \quad Z = \frac{\mu}{(ar^2 + 2d \cdot z + f)^{\frac{3}{2}}}.$$

Using again the equations (7.23) and (7.24), we can find two possible central forces satisfying the first two Kepler laws for motion:

$$F = \frac{\mu R c^{\frac{3}{2}}}{(bu + ev + c)^3} \quad \text{and} \quad F = \frac{\mu R}{(au^2 + 2d \cdot uv + fv^2)^{\frac{3}{2}}}. \quad (7.25)$$

These are also results obtained by Gaston Darboux on the same problem of finding the magnitude of central forces respecting the first and second of the Kepler laws [(Darboux, 1877, 1884); (Halphen, 1877)]. However, the Appell transformation bestows upon Halphen's solution a connotation of principle having profound consequences.

As we just said before, the notation (r, z) used here is intended to show a historical fact: the Appell transformation mimics the reduction of the general Newton theory to the Galileo's kinematics, a procedure which is known today as the reduction procedure, indeed (Carinena, Clemente-Gallardo, & Marmo, 2007). For our case here, the notation suggests a vertical plane projection of the horizontal dynamics, like in the case of tangent plane to Earth's surface. The horizontal motion represented by 'r' is uniform according to equation (7.24), while the vertical motion, represented by 'z' is uniformly accelerated, of course if the force Z is constant. This, indeed, carries out a projection, of what happens in the 'horizontal plane' containing the Kepler orbit described in coordinates (u, v), onto what happens in a 'vertical plane' in coordinates (r, z). As shown here, the projection can only be mathematically realized in terms of asymptotic directions to surface, which, by Appell's transformation, turn out to have a precise physical interpretation in terms of Newtonian central forces.

In order to physically explain the result we need to go back to Newton's Corollary 3 of the Proposition VII from the first book of Principia (Mazilu & Porumbreanu, 2018), assuming this time the existence of a physical ray issuing from a point located somewhere along a transversal direction to the plane of Kepler motion. The portion of any wave surface which may be taken to define this physical ray, can then be geometrically described in the parameters (u, v), whereby the Kepler orbit appears as the intersection of that wave surface with the plane of motion. So, the Kepler motion determines a wave surface, but only locally, in the manner in which the ellipse of polarization determines the wave surface in the case of light. Chances are, therefore, that this way one can generalize the Fresnel's synthesis of the wave surface from its differential elements, to a synthesis involving any kind of 'elements', *even non-differential fragments*, if we take into consideration the spirit of SRT. For the general idea of a synthesis of the wave surface, one of the Hamilton's works is to be recommended, as the clearest one from geometrical analytical point of view (Hamilton, 1841).

One more observation, before we go any further: we talked right above of 'asymptotic directions' and the involvement of these may not be quite so obvious for the casual reader. However, what we expressly had in mind at this juncture, is the fact that passing from the coordinates (u, v) to the coordinates (r, z) involves intuitively a transition between two scales of spacetime. Our contention is that the procedure, far from having only an intuitive meaning, like it had to Newton for instance, should actually be taken as a universal procedure and theorized as such. For, imagine a portion of the orbit of a satellite around the Earth, as in Newton's gedanken experiment with a cannon powerful enough, and high enough above ground, to launch such a satellite. This way one can prove indeed that the gravitation in the universe is the same one as the gravitation causing the free fall, provided one agrees to jump over the scale of things. For, the laws of Galilei for the free fall are local laws, valid over a finite altitude at the Earth surface, while the universality refers, obviously, to... universe. The above geometrical procedure transforming the Kepler motion in a free fall *in every point*, requires evidently a further physical explanation from the perspective of SRT. As it turns out, it is indeed legal, so to speak, but involves *precise formulations of scale transitions*.

The Three-Dimensional Space of Accelerations

So we come to the suggestion of using a piece of a surface as a reference for the surrounding space, exactly the way we use it daily in the case of Earth surface. The previous use of the Appell coordinates encourages us to hope that this intuitive behavior can be molded into a general theory with the help of idea of *wave surface*. However, choosing a wave surface element as reference for the surrounding space, leads to particular concerns regarding the metric quantities by which that space is mathematically turned into a concept. For instance, equation (7.11) shows that the height of the surface with respect to its tangent plane is given by a quadratic form involving some curvature parameters or their variations. This case is characteristic to the definition of second fundamental form of the surface, giving therefore positions of a point in the surrounding space along the normal to the surface element (Struik, 1988). Of course, this geometric interpretation resides upon the idea of infinitesimals, therefore upon a limiting process, but the theory can be extended into a representation of the height above surface by a Taylor series, truncated to the second order term, which is quite a general case.

Theoretically, one can argue that the dimension of the environmental space cannot be three, as classically suggested, but the issue is a little more involved, for it is not always purely geometrical. Of course, one cannot bring arguments against the intuitive feeling that the residence space of the matter, as it presents itself to our senses, is three-dimensional. The simplest argument for this statement is provided by the fact that in order to evaluate quantitatively the shape of a body, no matter of the space scale, we need at least three measurements of its space extension along three different directions. However, from the very foundation of the theory of surfaces — the linear independence of the vectors — the things cannot remain at the intuitive level, because the physics itself points out, in fact, to a environmental space of geometrical dimension five. Indeed, our line of introducing the physics here is based on the fact that both the metric form and the second fundamental form in a point of a surface are *binary quadratic forms*. The binary variable is provided, of course, by the components of differential displacements along the tangent plane to surface in the chosen point. It is from this perspective that the 'vertical' position of a certain point in the space referred to a surface element, along local normal direction to this surface is, in a sense, a vector in a three dimensional real space. The mathematical basis of this statement can be presented as follows.

Any binary quadratic form can be uniquely expressed as a linear combination of three nonsingular, mutually apolar, binary quadratic forms. The proof here is due to Dan Barbilian (Barbilian, 1971), but for the general theory of apolarity of the binary quadratic forms, one can consult a classical treatise (Burnside & Panton, 1960). The chapter XVI of the second volume of this last work, particularly the example 6, pp. 136-137, is particularly illuminating for what we have to say here. Start with designating a generic binary quadratic form by

$$Q(x, y) \equiv ax^2 + 2bxy + cy^2$$

where the binary variable is, again generically, denoted (x, y). Write this quadratic form as a linear combination

$$Q(x, y) \equiv \lambda_1 Q_1(x, y) + \lambda_2 Q_2(x, y) + \lambda_3 Q_3(x, y) \qquad (7.26)$$

where Q_k represent three quadratic forms whose coefficients are a, b, c with different indices 'k'. This identity is equivalent to the linear algebraical system:

$$\begin{aligned} a_1\lambda_1 + a_2\lambda_2 + a_3\lambda_3 &= a; \\ b_1\lambda_1 + b_2\lambda_2 + b_3\lambda_3 &= b; \\ c_1\lambda_1 + c_2\lambda_2 + c_3\lambda_3 &= c. \end{aligned} \qquad (7.27)$$

Considered as a linear system for the unknowns $\lambda_{1,2,3}$, this system can be compatible, and if the three basic quadratics are *nonsingular and mutually apolar*, it is always compatible and has unique solution. Indeed, the square of its third order principal determinant can be expressed as

$$\begin{vmatrix} a_1 & a_2 & a_3 \\ b_1 & b_2 & b_3 \\ c_1 & c_2 & c_3 \end{vmatrix}^2 = - \begin{vmatrix} I_{11} & I_{12} & I_{13} \\ I_{21} & I_{22} & I_{23} \\ I_{31} & I_{32} & I_{33} \end{vmatrix}; \quad 2I_{kl} \equiv a_k c_l + c_k a_l - 2b_k b_l.$$

$$(7.28)$$

Now, if the three basic quadratic forms are apolar $I_{kl} = 0$ for $k \neq l$, so that only the diagonal entries survive here, and these are the discriminants of the quadratic forms. Because they are assumed nonsingular, the equation (7.28) shows that the determinant of system (7.27) is always nonzero, therefore the system has *unique solution* for the unknowns $\lambda_{1,2,3}$. If the system is homogeneous, it has only the trivial null solution. Therefore the set of all quadratic binary forms, defined by some external reasons in a point of an affine surface, is *a linear three-dimensional set*.

Strange as it may seem, the dimension three for the 'normal space' to a physical surface has indeed strong physical reasons, not only historically as noted before, but even from the very classical mechanics' point of view. That point of view is given by the Newtonian relation between force and acceleration — the second law of dynamics — however, with a proper perspective of concepts. In the classical differential geometry of surfaces, as presented above, this perspective amounts to the following. Assume that $\mathbf{v}[B$ is the local speed of a motion along the geodesics of a surface in a certain point:

$$\mathbf{v} = \dot{s}^1 \hat{\mathbf{e}}_1 + \dot{s}^2 \hat{\mathbf{e}}_2.$$

Here the overdot denotes a time rate as usual. This is an intrinsic vector, just like the elementary displacement on the surface itself, from which it is derived. The rate of displacements is taken with respect to the time of geodesics of surface, which is to assume that they exist and the motion along them is physical. The variation of vector $\mathbf{v}$ along geodesics has only one component, namely that component normal to surface, according to the very definition of geodesics:

$$d\mathbf{v} = (\dot{s}^\alpha \Omega_\alpha^3) \hat{\mathbf{e}}_3.$$

Using the definition of the curvature vector, one can see that the quadratic generated by the second fundamental form, or its variation, in association with the time of geodesics on surface, i.e. the rate

$$\frac{dv}{dt} = \dot{s}^\alpha h_{\alpha\beta} \dot{s}^\beta \tag{7.29}$$

is the magnitude of a *normal acceleration to surface*. Here $\mathbf{h}$ is a general matrix, representing either the curvature matrix or its variation, as the case may occur. By the same token, the intrinsic vector normal to geodesics in the tangent plane

$$\mathbf{p} = \dot{s}^2 \hat{\mathbf{e}}_1 + (-\dot{s}^1) \hat{\mathbf{e}}_2$$

when transported by paralelism along geodesics, yields a normal vector

$$d\mathbf{p} = (\dot{s}^2 \Omega_1^3 - \dot{s}^1 \Omega_2^3) \hat{\mathbf{e}}_3 \tag{7.30}$$

126

having the magnitude

$$\frac{dp}{dt} = \dot{s}^\alpha h^*_{\alpha\beta} \dot{s}^\beta; \quad \mathbf{h}^* \equiv \mathbf{h} \cdot \mathbf{i}; \quad \mathbf{i} \equiv \begin{pmatrix} 0 & -1 \\ 1 & 0 \end{pmatrix}. \tag{7.31}$$

This is an acceleration due to the 'geodesic torsion': the vector (7.30) represents the torsion of any curve touching the geodesic in the given point. The two quadratic forms from equations (7.29) and (7.31), represent two *different accelerations*, both *oriented along the normal to surface*.

In a regular geometry these are just the magnitudes of two collinear vectors. Nevertheless, as quadratic forms, they are insufficient for characterizing the whole magnitude of the normal acceleration in general, inasmuch as, considered as a quadratic polynomial, this one is a point in a three-dimensional linear space, as shown above. This is the right place to use the property of apolarity in order to properly make up the magnitude of an acceleration. Indeed, as one can see directly, the two quadratics (7.29) and (7.31) are reciprocally apolar: the quantity I_{12} defined in equation (7.28), calculated with their coefficients, is null. Then we can naturally construct a third quadratic, apolar with each one of them, the so-called *resultant*:

$$\dot{s}^\alpha h^{**}_{\alpha\beta} \dot{s}^\beta; \quad \mathbf{h}^{**} \equiv \frac{1}{2}(h\mathbf{1} - \mathbf{h}^2); \quad \mathbf{1} \equiv \begin{pmatrix} 1 & 0 \\ 0 & 1 \end{pmatrix} \tag{7.32}$$

where 'h' is the determinant of $\mathbf{h}$, i.e. the Gaussian curvature of surface, or its variation. If the support function of the surface [see (Spivak, 1999) for a clear explanation of the concept] can be represented by a quadratic form, then the proper representation of the magnitude of acceleration, when referred to the time of the geodesics on the surface, should be by a linear combination of the quadratics (7.29), (7.31) and (7.32).

Therefore, a first general conclusion, concerning physics alone, can be drawn by assuming that by the classical Newtonian relation between the magnitude of acceleration and the magnitude of force we can transfer the reasoning upon forces. Then the forces with which the surface of a spatially extended particle responds when accomodating with each other the exterior structure to which the particle belongs physically, and interior changes of the particle's matter, should be linear combinations of the three accelerations above. One can further say, again on behalf of physics itself, that the coefficients entering such an expression of the forces represent *three types of inertia*, which only accidentally reduce to one, in those cases in which the force can be sufficiently characterized as vector. For instance, this could be the case when the constituent Hertz particles of the matter structure can be represented as material points. This was actually the very case that allowed introducing forces in physics.

Indeed, this was the manner of conceiving forces responsible for the celestial harmony in the first place: namely, by the ratio of their magnitudes (Newton,

Principia, Book I, Proposition VII, Corollary 3). Only, in the original case of Newton, the forces were considered as acting in different arbitrary directions in the plane of motion. A particular instance of such directions is given by the force from a planet toward the focus of its orbit, and from the same planet toward the center of the orbit. This provides the Newtonian force of magnitude inversely proportional with the square of the distance from the focus. However, a more explicit example of forces having as magnitude *a linear combination of magnitudes of forces*, is provided by *the case of revolving orbits*, whereby the magnitude of the force responsible for the motion along a revolving Kepler orbit is a linear combination of the magnitude of a force inversely proportional to the square of distance and the magnitude of a force inversely proportional to the cube of distance (Principia, Book I, Proposition XLIV). For a treatment of the problem in modern theoretical terms one can consult many works referring, for instance, to the basic reasons of general relativity. However, there are a few works considering the problem in its own classical right [see (Whittaker, 1917); (Chandrasekhar, 1995); (Lynden-Bell, 2006)].

There is, however, a very instructive classical case illustrating the issue even from the very standpoint of the theory of surfaces. Like anticipating the case of Fresnel, occurring two centuries later, Newton explained the phenomena of reflection and refraction of light by a force acting at the surface of matter, along the normal to that surface [see (Newton, 1952), pp. 79ff]. Nowadays, we should be able to say that such a force can be explained within differential geometry, by the previous extension of the principle of inertia, as long as the light itself is considered a material structure, as it always was actually. From this perspective, one can say that the diffraction phenomena studied by Fresnel, only extended the phenomenological framework leading to Newtonian force. The original phenomenology of Newton's optics was only based upon the experience regarding only *reflection* and *refraction of light*, while Fresnel enlarged this phenomenology with the experience on *diffraction*. The existence of the force of light as assumed by Newton, was indeed confirmed as a *physical fact*, however, only at *the infrafinite scale of time*, i.e. within a theory of collisions, on one hand through the electromagnetic structure of the light and, on the other hand, through the experiments of Pyotr Nikolaevich Lebedev [(Lebedeff, 1900); (Lebedew, 1902)]. The extension of the principle of inertia has, nevertheless, by far larger implication than the corpuscular theory of light.

Force at an Outward Distance

Assuming, indeed, that we have an *extended particle model*, the problem is to incorporate such a particle within a physical structure of the matter: that physical structure to which the particle belongs. A first issue in solving this problem, at least in a classical view, is the construction of a potential of the forces acting outside the matter of this particle, and we are able to show here that this potential is strictly determined by the limit of separation of the matter of extended particle. The expression of the force obtained from this potential depends solely on the curvature of surface and its variation. This can be shown as follows.

To start with, we take notice of the fact that by the beginning of the last century, Edmund Taylor Whittaker was able to produce solutions of the nonhomogeneous Laplace equation (in fact of Helmholtz equation) by introducing arbitrary functions in an integral over the unit sphere (Whittaker, 1903). Specifically, if we parametrize the sphere by the polar angles θ, ϕ with respect to its center, then we find that the potential written in the form

$$U(\mathbf{r}) = \oiint_{Sphere} e^{x \sin\theta \cos\phi + y \sin\theta \sin\phi + z \cos\theta} f(\theta, \phi) d\theta d\phi \qquad (7.33)$$

with $f(\theta, \phi)$ *an arbitrary function*, is a solution of the equation $\Delta U(r) = U(r)$, where Δ is the Laplacian in Cartesian coordinates (x, y, z), considered as components of the position vector $\mathbf{r}$.

Some three decades afterwards, Pierre Humbert noticed the fact that the partial differential equation to be satisfied by the potential of a surface in the outer space is actually *dictated by the equation of the surface itself*, if it can be represented by a known algebraical expression, as in the case of Whittaker's unit sphere. Such a description of the surface is always handy, even though far from unique, if one asks for means of defining it depending exclusively on the local geometry of surface. This can be shown quite directly, considering $U(\mathbf{r}, \mathbf{a})$ the potential between points $\mathbf{r}$ and $\mathbf{a}$ in space (Humbert, 1929). Humbert chose for illustration a Gaussian central potential, but he soon noticed that the theory goes very well with *arbitrary forms of potential*. It is this fact that entitles us to write the potential in terms of a simple algebraical approximation of the local elements of surface. The theory goes along the following lines: if the point of position $\mathbf{a}$ describes a surface, S say, the potential of that surface in the space point of position $\mathbf{r}$ is naturally given by the integral

$$U(\mathbf{r}) = \iint_S U(\mathbf{r}, \mathbf{a}) \, d^2\mathbf{a}$$

where $d^2\mathbf{a}$ is the elementary measure on that surface. For a potential of the form

$$U(\mathbf{r}, \mathbf{a}) = e^{\mathbf{a} \cdot \mathbf{r}} f(\mathbf{a}) \qquad (7.34)$$

and if the surface is algebraic, having the equation:

$$\sum_{m,n,p} a^m b^n c^p = 1 \qquad (7.35)$$

the potential as a function of position in space satisfies the partial differential equation:

$$\sum_{m,n,p} \partial_x^m \partial_y^n \partial_z^p \, U(\mathbf{r}) = U(\mathbf{r}). \tag{7.36}$$

These results generalize the one due to Whittaker for the representation of the solutions of Laplace equation: in that case it is sufficient to consider that the algebraic equation in (7.35) is a sphere of unit radius referred to its center, which can be obtained as a particular case for $m + n + p = 2$.

However, the Humbert's results can be seen even from another angle, namely as an 'eigenvalue problem', involving some 'level surfaces', so to speak. Indeed, if instead of (7.35) the equation of the local algebraical surface is of the general form

$$\sum_{m,n,p} A_{mnp} a^m b^n c^p = \lambda \tag{7.37}$$

where A_{mnp} and λ are some parameters, then the partial differential equation satisfied by $U(\mathbf{r})$, as given by equation (7.34), is of the form:

$$\sum_{m,n,p} A_{mnp} \partial_x^m \partial_y^n \partial_z^p \, U(\mathbf{r}) = \lambda U(\mathbf{r}). \tag{7.38}$$

This allows us to include among the representations of Humbert some important homogeneous equations like, for instance, the Laplace equation proper, which is the epitome of the classical theory of potential. The solutions of this equation are then represented by an integral of the form

$$U(\mathbf{r}) = \iint_S e^{\mathbf{r} \cdot \mathbf{a}} f(\mathbf{a}) \, d^2 \mathbf{a} \tag{7.39}$$

where $f(\mathbf{a})$ is an arbitrary function on the surface $\mathbf{a}^2 = 0$ [see also (Bateman, 1918), Chapter VIII, and (Helgason, 1984), pp. 18 - 20]. Thus, in view of what has been said before, the potential of an extended material particle in a position from the outer space, considered as a solution of the Laplace equation in that position, is decided by what we like to call *the spin parameters* of the extended particle (Yamamoto, 1952). If these spin parameters are connected to the curvatures parameters and their variations, the force generated by an extended particle in any position from the outside space is offered by the gradient of the potential, as usual:

$$\nabla U(\mathbf{r}) = \iint_S \mathbf{a} e^{\mathbf{r} \cdot \mathbf{a}} f(\mathbf{a}) \, d^2 \mathbf{a} \tag{7.40}$$

a conclusion in concordance with the extended inertia principle as restated above. The arbitrary function $f(\mathbf{a})$ can be given from further invariance considerations, *but related to surface only.*

It is now the proper place to take notice of the inedit: the *inertia property* of matter is an issue involving the surface separating the matter form space, which means that *it should be a holographic property*. Indeed, the three spin parameters as characterized by Yamamoto, can be realized as accelerations as we have shown above. Thus, they occur with a proper physical characterization of the surface of a certain material point in the sense of Hertz. The force of inertia is thereby related to this surface, and from it any external force can be calculated by the recipe given in equation (7.40). This equation is then the universal expression of the second principle of dynamics, with a proper definition of the surface forces.

Chapter 8. Nonconstant Curvature

Consider, therefore, the case of a nonconstant curvature matrix, and choose the variations of curvature as determining the variations in the direction of the normal to surface:

$$|d\mathbf{\Omega}^3\rangle = d\mathbf{b} \cdot |s\rangle$$

as an equation for the 'orbits' through $|s\rangle$, as it were. This is surely the case with a special gauging of the local piece of surface, with the assist of curvature matrix itself. If, for instance, the curvature is initially decided by a Kepler orbit, as in equation (6.38), and the surface is locally gauged by this Kepler orbit as in equation (6.39), then the curvature vector itself satisfies to a known differential equation involving the coframe of the SL(2, R) algebra:

$$|d\mathbf{\Omega}^3\rangle = (d\mathbf{b} \cdot \mathbf{b}^{-1}) \cdot |\mathbf{\Omega}^3\rangle. \tag{8.1}$$

Using the three curvature parameters, α, β, γ say, the equation (8.1) is:

$$|d\mathbf{\Omega}^3\rangle = (d\lambda \cdot \mathbf{1} - \mathbf{\Omega}) \cdot |\Omega^3\rangle \tag{8.2}$$

where we denoted:

$$\lambda = ln\sqrt{\alpha\gamma - \beta^2}; \quad \mathbf{\Omega} = \begin{pmatrix} \omega^2/2 & -\omega^1 \\ \omega^3 & -\omega^2/2 \end{pmatrix} \tag{8.3}$$

with $\omega^{1,2,3}$ the differential 1-forms of the coframe:

$$\omega^1 = \frac{\alpha d\beta - \beta d\alpha}{\alpha\gamma - \beta^2}; \quad \omega^2 = \frac{\alpha d\gamma - \gamma d\alpha}{\alpha\gamma - \beta^2}; \quad \omega^3 = \frac{\beta d\gamma - \gamma d\beta}{\alpha\gamma - \beta^2}. \tag{8.4}$$

Now, in order to interpret geometrically, and therefore physically, this definition, we need first to have some characterization of the vector from equation (8.2) with respect to the position $|s\rangle$ from the tangent plane. This will then show how much the second fundamental form differs from the one calculated in the position of origin of the plane. From equation (8.2) we have by direct dot multiplication:

$$\langle s|d\mathbf{\Omega}^3\rangle = (d\lambda)\langle s|\Omega^3\rangle - \langle s|\mathbf{\Omega}|\Omega^3\rangle \tag{8.5}$$

where the matrix $\boldsymbol{\Omega}$ is defined in equation (8.3). Therefore the variation 'gauged' through equation (8.2) induces a differential variation along the position vector in the tangent plane proportional with the second fundamental form, but to which a certain quadratic form is added. Using equation (7.9) for calculating this quadratic form, results in

$$\langle s|\boldsymbol{\Omega}|\Omega^3\rangle \equiv \langle s|\boldsymbol{\Omega}\cdot\mathbf{b}|s\rangle$$
$$= \left(\frac{\alpha}{2}\omega^2 - \beta\omega^1\right)(s^1)^2 + (\alpha\omega^3 - \gamma\omega^1)(s^1)^2 \qquad (8.6)$$
$$+ \left(-\frac{\gamma}{2}\omega^2 + \beta\omega^3\right)(s^2)^2.$$

The general conclusion is that the variation of curvature parameters at a given position *on the surface* involves three quadratic differential forms: the second fundamental form of surface in the given position, the quadratic differential from equation (8.6) and the quadratic differential

$$-\langle d\Omega^3|^*s\rangle = \omega^1(s^1)^2 + \omega^2 s^1 s^2 + \omega^3(s^2)^2 \qquad (8.7)$$

where the star denotes the usual Hodge duality operation on the differentials: $|^*s\rangle \equiv \mathbf{i}\cdot|s\rangle$. Now, the two differential forms involved in equation (8.5), together with the one from equation (8.7) are, from algebraic point of view, a system of three *mutually apolar quadratics*: the very property of the spin parameters in Yamamoto representation. The idea is thereby suggested, that the apolarity plays a fundamental role, when it comes to issues of the variation of curvature, and consequently to the physics involved in this variation of curvature.

Thus, as long as we refer the space to a surface, there are always three mutually apolar quadratics that can play the part of spin parameters, and these are generated by the variations of curvature of the surface. These quadratics may play the parts of some accelerations as in the case of the classical theory of surfaces, equations (7.29), (7.31) and (7.32), but there are some problems. The first of these is the calibration condition leading to the equation (8.1): it requires a condition of calibration like (6.39), and if there are forces involved here – as there should be, if we are talking about the Kepler problem – these need to satisfy the *Wigner dynamical principle*, not the classical second principle of dynamics. Thus, in the general formulation of the dynamical principle we need to account for *calibration forces* setting the background, so to speak, and *incidental forces*, responsible for the variation of curvature *per se*, and acting on that background. A case of working in such a background should be illuminating here.

The Infinitesimal Deformation

In order to give an interpretation of the previous results we shall call upon one of the many descriptions of the deformation of surfaces, namely the *infinitesimal deformation* [see (Guggenheimer, 1977), p. 245ff]. Such a deformation of a certain surface $\mathbf{x}$ gives a new surface described by a position vector $\mathbf{r}$ of the form:

$$\mathbf{r}(\varepsilon) = \mathbf{x} + \varepsilon \cdot \mathbf{z}.$$

The choice of symbols is here intended to show that the surface $\mathbf{x}$ undergoes a deformation characterized by the small parameter ε, in a 'vertical direction', so to speak, given by vector $\mathbf{z}$. In general, however, the direction of deformation is not vertical *per se*, i.e. it is not normal, but only transversal to the surface $\mathbf{x}$, in the sense that it takes the corresponding points out of this surface. By definition, this type of deformation is such that the variation of the first fundamental form is null for small ε, i.e. it remains approximately constant through the deformation:

$$\lim_{\varepsilon \to 0} \frac{d\mathbf{r} \cdot d\mathbf{r} - d\mathbf{x} \cdot d\mathbf{x}}{\varepsilon} = 0 \tag{8.8}$$

where the geometry of the ambient space is considered to be Euclidean. This condition leads to the following constraint for the vector $\mathbf{z}$:

$$d\mathbf{x} \cdot d\mathbf{z} = 0 \quad \therefore \quad d\mathbf{z} = \mathbf{y} \times d\mathbf{x}. \tag{8.9}$$

Here $\mathbf{y}$ is an ancillary vector which, nonetheless, cannot be quite arbitrary. Its properties are, again, simple consequences of the exterior differential calculus. Indeed, the vector $d\mathbf{z}$ in equation (8.9) should be an exact differential vector. Imposing this condition on the second of these equations, and using again Cartan's Lemma 1, leads directly to the following equations:

$$v^1 = Bs^1 + Cs^2, \quad -v^2 = As^1 + Bs^2, \quad v^3 = 0 \tag{8.10}$$

where the definition $\mathbf{v} \equiv d\mathbf{y}$ is used. Thus, the ket $|v\rangle$, as defined by this last equation, is practically a vector *intrinsic to surface*, which in turn means that the normal component of the ket $|y\rangle$ itself must be, locally, a constant. Moreover, by its definition $|v\rangle$ must be an exact differential vector. The condition that its exterior differential should be the null vector, leads to equations like (7.7), but with v^1 and v^2 instead of s^1 and s^2 . The third of those equations, combined with equation (8.10), then shows that A, B, C are constrained to satisfy the linear relation

$$\alpha C + \gamma A - 2\beta B = 0 \tag{8.11}$$

where α, β, γ are the coefficients of the second fundamental form (the entries of the curvature matrix). This shows that the flux through surface at the chosen location, given by equation:

$$d\mathbf{x} \times \mathbf{v} = \langle s|\mathbf{i}|v\rangle \hat{\mathbf{e}}_3; \quad \langle s|\mathbf{i}|v\rangle \equiv A(s^1)^2 + 2Bs^1s^2 + C(s^2)^2 \quad (8.12)$$

where $\mathbf{i}$ is the fundamental skew-symmetric 2×2 matrix, has as magnitude a quadratic form which is apolar to the second fundamental form. It only remains to elaborate on how the coefficients A, B, C can be related to the physics of our problem.

Notice that the parameters A, B, C are, again, external parameters, introduced by Cartan's Lemma 1. They can be introduced as such, in the way in which the curvature parameters were introduced, as proceeded above, or they can be constructed from the curvature parameters. This last case is always handy by the very rules of algebra. Indeed, the equation (8.11) is an apolarity condition, between the quadratic form (8.12) and the second fundamental form of the surface in the point considered. If we take these parameters as differential forms (8.4), according to the rule of correspondence

$$A \leftrightarrow \omega^1, \quad 2B \leftrightarrow \omega^2, \quad C \leftrightarrow \omega^3 \quad (8.13)$$

we can define an infinitesimal deformation, induced exclusively by the variation of curvature parameters. This infinitesimal deformation is then expressed by an addition to the second fundamental form, as we already mentioned before, given by equation (8.12) as:

$$\langle s|\mathbf{i}|v\rangle \equiv \omega^1(s^1)^2 + \omega^2 s^1 s^2 + \omega^3(s^3)^2. \quad (8.14)$$

This vector has indeed the magnitude given by equation (8.7), and therefore can be viewed as related somehow to the ket $|d\Omega^3\rangle$ defined in equation (8.1). It can be described indeed by an auxiliary ket $|y\rangle$ defining a deformation, which is directed along the normal to surface, and which acquires in-surface components by the very curvature parameters' variations:

$$\begin{pmatrix} 0 \\ 0 \\ y^3 \end{pmatrix} \rightarrow \begin{pmatrix} (\omega^2/2)s^1 + \omega^3 s^2 \\ -\omega^1 s^1 - (\omega^2/2)s^2 \\ y^3 \end{pmatrix}. \quad (8.15)$$

The corresponding first fundamental form then becomes:

$$\langle s|\mathbf{h}|s\rangle; \quad \mathbf{h} = [1 + (\varepsilon y^3)^2] \cdot \mathbf{1} + \varepsilon^2 (\mathbf{i} \cdot |v\rangle\langle v| \cdot \mathbf{i}) \quad (8.16)$$

where the ket $\mathbf{i} \cdot |v\rangle$ is calculated based on equation (8.10), but this time with the values (8.13) for the coefficients A, B, C. This matrix obviously reduces to the usual identity matrix, representing the initial Euclidean geometry, once the parameter $\varepsilon \to 0$.

We can assume, therefore, that the initial surface generated on the basis of a Kepler motion, is an already deformed surface, and thus start in the description of the process of deformation from an initial surface having a nontrivial metric. Then, the physics of this problem reveals another way of construction of the infinitesimal deformation, whereby the classical Euclidean notions enter explicitly. Specifically what Euclidean notions, is quite obvious from the dynamical theory of Kepler orbits. Namely, such an orbit can be identified by the initial conditions from which it starts. Thus, if the orbiting matter is to be physically interpreted in the manner of Louis de Broglie, and the interpretation is to be achieved by ensembles of Hertz material particles, these particles can be identified only by their velocities. And the physically fundamental problem is that of constructing a surface which is perpendicular to such a velocity vector. Of course, as long as only the Euclidean geometry is involved, we just have to apply a procedure based on relations (7.10) and (7.11). The problem is a little more complicated when we already have a surface at our disposal, like, for instance, the one determined by an isolated Kepler orbit. In this case the curvature parameters are given, and based on these parameters we need to find a surface normal to the velocity field. Then the following general recipe can be applied.

Assume a certain Euclidean velocity field $\mathbf{v}$, and a surface locally characterized by the curvature parameters $\alpha^1 = \alpha, \alpha^2 = \beta, \alpha^3 = \gamma$. This last indicial notation entails a warning about the procedure we are just about to describe: they are the contravariant components of some position vector locally describing the surface, as shown before, for instance the coefficients of an ellipse considered as Dupin indicatrix of the surface. Case in point, the ω^k from equation (8.4) might be considered as some vector densities constructed with a metric given by the discriminant of the second fundamental form. Likewise, the components of vector $\mathbf{v}$ are the coefficients of a differential form in space: $v = v_k \cdot dx^k$, i.e. some 'covariant' components, carrying lower indices. If the vectors involving the components α^k are a realization of the structure relations of the three- dimensional $\mathbf{sl}(2,\mathrm{R})$ algebra, satisfying the structure given by equations (2.43), viz.:

$$[\mathbf{X}_i, \mathbf{X}_j] = C_{ij}^k \mathbf{X}_k \qquad (8.17)$$

then we can construct the bilinear forms

$$\xi_i \equiv \alpha^j C_{ij}^k v_k \equiv \langle \alpha | \mathbf{h}_i | v \rangle. \qquad (8.18)$$

One has to understand here that an upper index represents the number of a line of the matrix $\mathbf{h}_i$, defined by

$$(\mathbf{h}_i)_j^k \equiv C_{ij}^k \qquad (8.19)$$

136

so that, according to equation (2.43), we will take:

$$\mathbf{h}_1 = \begin{pmatrix} 0 & 1 & 0 \\ 0 & 0 & 2 \\ 0 & 0 & 0 \end{pmatrix}; \ \mathbf{h}_2 = \begin{pmatrix} -1 & 0 & 0 \\ 0 & 0 & 0 \\ 0 & 0 & 1 \end{pmatrix}; \ \mathbf{h}_3 = \begin{pmatrix} 0 & 0 & 0 \\ -2 & 0 & 0 \\ 0 & -1 & 0 \end{pmatrix}. \quad (8.20)$$

Thus, the matrices $\mathbf{h}_k$ act upon α's by left multiplication, and on v's by right multiplication, and thus we have, from equation (8.18):

$$\xi_1 = \alpha^1 v_2 + 2\alpha^2 v_3; \ \xi_2 = -\alpha^1 v_1 + \alpha^3 v_3; \ \xi_3 = -2\alpha^2 v_1 - \alpha^3 v_2. \quad (8.21)$$

These three quantities satisfy two important identities, namely:

$$\xi_1 v_1 + \xi_2 v_2 + \xi_3 v_3 = 0; \quad \xi_1 \alpha^3 - 2\xi_2 \alpha^2 + \xi_3 \alpha^1 = 0 \quad\quad (8.22)$$

which may be able to justify us in calling them 'transition quantities'. Indeed, they play a double role: on one hand the ξ_k can be considered as the components of an Euclidean vector perpendicular to vector $\mathbf{v}$, while, on the other hand, they can be considered as the coefficients of a quadratic form, apolar to the second fundamental form of the surface characterized by α^k .

Perhaps this description is not enough for understanding the idea of transition between the two otherwise intransitive geometries. However, let us think of the fact that we are always prepared to accept the notion that the *matter fills a space*. With Newton and, closer to our times, with Einstein, this notion became critical. And this 'criticalness', so to speak, can simply be exhibited as the common denominator of the classical physics and theory of relativity: *we do not know anything about the space filled by matter. All we know is referring to the matter* filling that space, the space itself is a problem of hypothesis. As we have previously shown, the idea of ensemble is intrinsically involved in the possibility of interpretation of matter *per se*. Along this interpretation the concept of surface comes as only natural with the fundamental model of our knowledge: the planetary model. This model shows that the $\mathbf{sl}(2,\mathbb{R})$ differential geometry is the *natural Riemannian geometry of matter*. And having some quantities characteristic to this geometry, which belong also to the Euclidean geometry, can be taken as an indication that the geometry of the space free of matter is, or at least can be taken as, Euclidean. It is then only natural to assume that the matter fills an Euclidean space, and work with this assumption along the ideas of Élie Cartan in constructing the Riemannian geometry of space based on the *concept of torsion* (Cartan, 1931)

Summing up the Differential Geometry of Curvature Parameters

A few algebraical relations among the differential forms from equation (8.4) are in order. They form a basis (coframe) of a SL(2,R) algebra. We already alluded to such a structure, by presenting equation (8.1) as a gauge equation obtained by left multiplication with the inverse of the curvature matrix. More than this, it turns out that, among other things, the space of curvature parameters can be organized as a Riemannian space. In order to show this, notice first the following differential relations in the space of curvature parameters, which can be proved by a direct calculation:

$$d \wedge \omega^1 = \frac{\alpha}{\sqrt{\Delta}}\Theta; \quad d \wedge \omega^2 = \frac{2\beta}{\sqrt{\Delta}}\Theta; \quad d \wedge \omega^3 = \frac{\gamma}{\sqrt{\Delta}}\Theta. \qquad (8.23)$$

Here Θ is the differential 2-form of Hannay

$$\Theta \equiv \frac{\alpha d\beta \wedge d\gamma + \beta d\gamma \wedge d\alpha + \gamma d\alpha \wedge d\beta}{\Delta^{3/2}}. \qquad (8.24)$$

The 2-form Θ is closed because it is the exterior differential of a 1-form:

$$\Theta \equiv d \wedge \psi; \quad \psi \equiv \frac{\alpha + \gamma}{\sqrt{\Delta}} d\left(\tan^{-1}\frac{2\beta}{\alpha - \gamma}\right)$$

which represents the classical *Hannay angle* for the problem of variation of the second fundamental form of a surface. In the present context this 1-form can be written as

$$\Psi = 2\frac{\omega^1 + \omega^3}{\sinh \xi}, \quad \xi \equiv \ln\frac{\alpha + \gamma}{2\sqrt{\Delta}} \qquad (8.25)$$

including explicitly the ratio between the mean and Gaussian curvatures. It gives therefore a way to establish the mathematical procedure for the local problem of surface forces, but certainly has everything in common with the original angle designated as such [see (Hannay, 1985); (Berry, 1985)]. More than this, there is also a statistic involved in this exterior calculus, which will be revealed shortly.

Meanwhile, continuing with pure algebra, we can verify the following relations:

$$\omega^1 \wedge \Omega^2 = \frac{\alpha}{\sqrt{\Delta}}\Theta; \quad \omega^2 \wedge \Omega^3 = \frac{\gamma}{\sqrt{\Delta}}\Theta; \quad \omega^3 \wedge \Omega^1 = -\frac{\beta}{\sqrt{\Delta}}\Theta. \qquad (8.26)$$

Thus, from (8.17) and (8.22) we have indeed the characteristic equations of a **sl(2, R)** structure given before in equation (2.49):

$$d \wedge \omega^1 - \omega^1 \wedge \omega^2 = 0;$$
$$d \wedge \omega^2 + 2(\omega^3 \wedge \omega^1) = 0; \qquad (8.27)$$
$$d \wedge \omega^3 - \omega^2 \wedge \omega^3 = 0.$$

Using these relations we can draw an important conclusion, destined to guide our future research. Notice, indeed, that in a given point of the surface, we can construct the 2-form in the curvature parameters' space:

$$\langle s|d \wedge (\mathbf{\Omega} \cdot \mathbf{i})|s\rangle = \frac{\langle s|\mathbf{b}|s\rangle}{\sqrt{\Delta}}\Theta; \quad \mathbf{i} \equiv \begin{pmatrix} 0 & -1 \\ 1 & 0 \end{pmatrix} \quad\quad (8.28)$$

where we have used the equation (8.17) and the matrix $\mathbf{\Omega}$ from equation (8.3). As the 2-form Θ is a flux in the space of curvature parameters, to wit something analogous of the solid angle in the usual Euclidean space, the second fundamental form $\langle s|\mathbf{b}|s\rangle$ of a surface, in a given point, is in fact the intensity (the value) of a flux in the space of curvature parameters, depending quadratically on the position in the local tangent plane. This philosophy can be profitably used in constructing a statistic of the fluxes of forces for a material point inside matter. First, however, let us define the notion of *surface tension*, in order to reveal the right place of the Riemannian character of the space of curvature parameters: its metric.

A Definition of Surface Tension

We introduce the surface tension locally, as a differential 2-form. It can be the component of a vector, or even the magnitude of a vector, depending on the measure of the deformation of surface we manage to define. Talking of deformation, it is the reason we consider the surface tension as being represented by a 2-form. Indeed, in order to interpret the deformation in the physical way, i.e. *by an ensemble in the continuous matter*, even from the times of Augustin Cauchy physics took the habit of considering the fluxes of forces between some constitutive material particles — called molecules in the times of Cauchy — through a plane. The deformation is then connected to the tensions created by these fluxes of forces calculated with respect to planes in matter, and designated as *stresses*, via a *constitutive law* of matter. Classically, one defines the stresses as tensions on the six faces of a little cube, imagined as cut out from the matter. The stresses are then the components of the resultants of these fluxes of forces through each one of the six faces of cube. With the present work, we have in mind two steps in generalization of this classical image. First, a little cube is quite a particular solid shape, which might indeed be assumed by a material point in matter, but this is highly unlikely to be realized as such. It is true that, inside matter, such a solid shape can indeed be imagined, but it rather counts only as a *reference frame*, as we explained before. For, inside matter, the ephemeral shapes of the lumps of matter assumed by a material point can be rather complicated, like higher dimensional polytopes, to say the least [see (Manton & Sutcliffe, 2004) for the gist of a description of material formations by such shapes].

139

It is then obvious that a material point has to be described only locally on *incidental surfaces*, therefore on incidental planes of arbitrary orientation, and on these planes we need to connect the tensions with an already built-in statistics somehow related to the deformation. So, as a first step in introducing physics here, we define the *local tension* as the 2-form:

$$f_0 = \Omega_1^3 \wedge \phi^1 + \Omega_2^3 \wedge \phi^2 \qquad (8.29)$$

where ϕ^1 and ϕ^2 are two *conveniently chosen* differential 1-forms, and the components of the differential of curvature vector are defined in equation (8.2). The equation (8.29) is the expression of a specific logic, allowed, this time, by using the *Cartan's Lemma 2*. Indeed, it incorporates the idea that the tension is directly related to curvature. For, according to Lemma 2, the tension f_0 is null whenever there is no curvature in the given point, and reciprocally, of course: if the tension is zero, there is no curvature. However, this tension obviously depends on some other physical circumstances, embodied this time in the choice of differential 1-forms ϕ^1 and ϕ^2 . This allows us to state that the tension can be zero even if there is curvature of the surface, as it happens indeed, in actual cases of our experience.

Thus, if the conveniently chosen auxiliary forms, in the definition (8.29) of the tension, are the components of the *first fundamental form*, i.e. the local components of the position vector $|s\rangle$ in the tangent plane for instance, then equation (8.29) simply shows that there is no tension. For then, there is a symmetric matrix relating $|\Omega^3\rangle$ to $|s\rangle$, as in equation (7.9), and therefore the tension is zero. On the other hand, if the tension is zero, according to Lemma 1 there is, again, a symmetric matrix defining the local curvature. Thus, the geometrical definition of the local curvature of a surface is simply a consequence of dynamics: *it represents a description of the local geometry in the absence of tensions on surface*. We can think of this definition as being the definition of a *local reference state* of the surface, with the state defined by the curvature parameters, viz. the entries of the curvature matrix **b**: a given set of values of these three parameters represents a fragment of surface, which represents geometrically *a state of the surface*.

Starting from such a state, we can build an evolution based on tensions, along the very same line of thought, but based on *Agaoka's theorem*. The tension, if it exists, should be related to the variation of curvature. We take a differential 3-form, assuming that it can be zero whenever two differential 2-forms are zero and vice versa. Then we can write:

$$f = f_1 \wedge \phi^1 + f_2 \wedge \phi^2. \qquad (8.30)$$

Here 'f' is our 3-form, and (f_1, f_2) are some 2-forms, components of the surface tension. Then, if the 3-form 'f' is null, we have, according to Agaoka's theorem:

$$|f\rangle = \mathbf{B} \wedge |\phi\rangle \qquad (8.31)$$

140

where **B** is a symmetric matrix having some 1-forms as entries. In the right hand side here, we have a matrix multiplication as usual, but with exterior product of 1-forms instead of usual multiplication of numbers. Now, we can assume that the ket $|\phi\rangle$ is generated by variation of the curvature parameters.

An digression is in order here: as one can see, the equation (8.30) cannot refer but to a three-dimensional theory, since the intrinsic theory of surfaces cannot accommodate an exterior 3-form. However, an embeding theory of surfaces can certainly contain such a form, which, in this case, can even be considered as mandatory from the point of view of interpretation. Fact is that in a Hertzian natural philosophy of matter we have to deal with two types of surfaces: one separating the categories of matter from the categories of space, the other delimiting the categories of matter themselves, specifically, material points, with respect to each other. This last type of surfaces, *accidental inside matter*, is the one serving in clarification of Madelung's notion of 'penetrating without merging'. 'Merging' is — at least in the framework of the classical natural philosophy, if not all over the natural philosophy in general — a notion related to the first type of surfaces, i.e. those describing the separation of matter from space. It involves forces calculated according to a Whittaker-type theorem, exclusively through the 'holographic properties', presented by matter in its limit of separation from space, as in equation (7.40), for instance. Therefore, the equation (8.30) represents a *matter surface embedded in matter*, thus 'penetrating' but not 'merging'. In hindsight, one can say that this property of the matter, so aptly described by Madelung in intuitive terms, has a significant history within natural philosophy. This history is epitomized by the idea which triggered the birth of relativity, that the *bodies move through ether without dragging it along*. At the time when this idea reached its critical point, and, in fact, even before, starting from its birth as a concept, the *ether was a category of matter*. So the first ever case of 'penetrating without merging' is simply the motion of bodies through ether.

Returning to our algebraical development, a well-known choice of the ket $|\phi\rangle$, describing the state of surface as defined above, and therefore strictly related to the **sl**(2,R) group algebra is *the Sasaki ket*, $|S\rangle$ say, having the components (Sasaki, 1979):

$$\phi^1 \equiv \omega^2; \quad \phi^2 = \omega^1 - \omega^3 \tag{8.32}$$

where $(\omega^1, \omega^2, \omega^3)$ is the coframe of this algebra, given in equation (8.4). The symmetric matrix $\mathbf{B}_0$, defined by

$$\mathbf{B}_0 \equiv \begin{pmatrix} \omega^1 & \omega^2/2 \\ \omega^2/2 & \omega^3 \end{pmatrix} \tag{8.33}$$

acts on this vector as an exterior differentiation:

$$\mathbf{B}_0 \wedge |S\rangle \equiv -\frac{1}{2}\mathbf{i} \cdot d \wedge |S\rangle; \quad \mathbf{i} \equiv \begin{pmatrix} 0 & -1 \\ 1 & 0 \end{pmatrix} \tag{8.34}$$

141

with **i** denoting the fundamental skew-symmetric 2×2 matrix.

The Statistics of Fluxes on a Material Point

The characterization of a local flux of forces is closely related to a plane centric affine geometry. That is to say that if one insists in characterizing a statistics of the contact forces on the surface of a certain nucleon, for instance, one has to consider the centric affine geometry in the tangent plane. This section shows a way to build such a statistics, based on the idea of continous Lie group characterizing the plane centric affine geometry.

The **sl**(2,R) action preserving origin of this plane geometry is given by the three vectors

$$X_1 = s^2 \frac{\partial}{\partial s^1}; \quad X_2 = \frac{1}{2}\left(s^1 \frac{\partial}{\partial s^1} - s^2 \frac{\partial}{\partial s^2}\right) \quad X_3 = -s^1 \frac{\partial}{\partial s^2} \quad (8.35)$$

while the corresponding action in the space of curvature parameters is realized by the vectors:

$$A_1 = -\alpha \frac{\partial}{\partial \beta} - 2\beta \frac{\partial}{\partial \gamma};$$

$$A_2 = -\alpha \frac{\partial}{\partial \alpha} + \gamma \frac{\partial}{\partial \gamma}; \quad (8.36)$$

$$A_3 = 2\beta \frac{\partial}{\partial \alpha} + \gamma \frac{\partial}{\partial \beta},$$

This last realization characterizes an intransitive action in the space of curvature parameters, which allows transitivity only along specific manifolds, given by constant discriminant of the second fundamental form. Therefore the action realized by operators (8.30) is transitive only at constant Gaussian curvature.

The functions of physical interest can be presented here as *joint invariants* of any two of the actions given by equations (8.35) and (8.36), with the help of *Stoka theorem* (Stoka, 1968). According to this theorem, any joint invariant of the two actions is an arbitrary continuous function of the two algebraic formations

$$\alpha(s^1)^2 + 2\beta s^1 s^2 + \gamma(s^2)^2, \quad \alpha\gamma - \beta^2. \quad (8.37)$$

Obviously the Gaussian probability density, for instance, if the case may occur, is only a special case of this theorem. By the same token, the straight lines through origin $s^1 = s^2 = 0$ can be presented as joint invariants of two actions realized by operators (8.35), while the joint invariants of two actions realized by operators (8.36), one in the variables a, b, c, say, the other in the curvature parameters α, β, γ, are arbitrary continuous functions of the following three algebraic formations (Mazilu, 2006):

$$\alpha\gamma - \beta^2, \quad ac - b^2, \quad a\gamma + c\alpha - 2b\beta. \quad (8.38)$$

These are important in problems transcending the manifolds of transitivity, of which an example will be given presently. These algebraic facts can give good reasons for a few further observations related to the classical statistical theory of contact points at the surface of a nucleon inside nucleus.

Before entering the calculational detail, let us notice that such a line of thought tips us to ammend the definition of a shape as given, for instance, by Shapere and Wilczek [see the works included in the collection (Shapere & Wilczek, 1989)]. Namely, we consider the instant shape, of a nucleon say, first of all as a collection of elementary events, described by contact points, their extended contact spots and the contact forces on them. It is this collection that should be considered as an evolving part of a 'phase space' of shapes. The actual space shapes have yet to be constructed from these elements by a certain physical procedure. The classical illuminating example is Fresnel's construction of the wave surface from pieces accessible to diffraction experiments.

Thus, for instance, consider that the fluxes of contact spots of a certain nucleonic surface, are controlled by the Dupin indicatrices at the contact points. According to Stoka theorem, the statistical ensemble of these contacts *may be characterized* by a normal probability density

$$
\begin{aligned}
& p_{XY}\left(s^1, s^2 | \alpha, \beta, \gamma\right) \\
& \equiv \frac{\sqrt{\alpha\gamma - \beta^2}}{2\pi} \exp\left\{ -\frac{1}{2}[\alpha(s^1)^2 + 2\beta s^1 s^2 + \gamma(s^2)^2] \right\}
\end{aligned}
\tag{8.39}
$$

in two statistical variables X and Y, of which we don't know too much for now, other than that they are the coordinates of position on any one of the contact spots of the surface of nucleon, as suggested before.

We have, therefore, a way to calculate the statistics of a quadratic variable Z(X, Y), obtained as before, in the process of deformation by contact, and having the generic values:

$$
z(s^1, s^2) \equiv \frac{1}{2}\left[a(s^1)^2 + 2bs^1 s^2 + c(s^2)^2\right].
\tag{8.40}
$$

Thus we need to find first the probability density of this variable, under condition that the plane of contact is characterized by the a priori probability density as given, for instance, by the Gaussian in equation (8.39). The probability density of Z should also satisfy the Stoka theorem, in the precise sense that it must be a function of the algebraic formations from equation (8.38). This leaves us with a functionally undetermined probability density though, even if we impose some natural constraints in order to construct it.

Nevertheless, proceeding directly, in the usual manner of the statistical theoretical practice, we are able to solve the problem, at least in this particular case. Thus, we have to find first the characteristic function of the variable (8.40). This is the expectation of the imaginary exponential of Z, using (8.39) as probability density. Performing this operation directly, we get, with an obvious notation for the average:

$$\langle e^{i\zeta Z}\rangle = \frac{1}{2\pi\sqrt{1 + (i\zeta)\dfrac{a\gamma + c\alpha - 2b\beta}{ac - b^2} + (i\zeta)^2\dfrac{\alpha\gamma - \beta^2}{ac - b^2}}}. \qquad (8.41)$$

In view of (8.38), this characteristic function certainly satisfies the Stoka theorem, which thus reveals its right place in a physical theory: it should serve for the selection of the right physical functions, specifically the probability density, or the characteristic function, as in this case. Anyway, the sought for probability density can then be found by a routine Fourier inversion of the characteristic function from equation (8.41), based on existing tabulated formulas [see (Gradshteyn & Ryzhik, 2007), especially the examples 3.384(43); 6.611 (40); 9.215(16) & (39)]. The result is:

$$p_z(z|\alpha, \beta, \gamma) = \sqrt{AB}\exp\left(-\frac{A + B}{2}z\right)\cdot I_0\left(\frac{A - B}{2}z\right). \qquad (8.42)$$

Here I_0 is the modified Bessel function of first kind and order zero, and A, B are two constants to be calculated from the formulas

$$A + B = \frac{2b\beta - a\gamma - c\alpha}{ac - b^2}; \quad AB = \frac{\alpha\gamma - \beta^2}{ac - b^2}; \quad A > B. \qquad (8.43)$$

Again, this probability density obviously satisfies the Stoka theorem, as it is a function of the joint invariants from equation (8.38). And so do the statistics of the variable Z, i.e. its mean and variance, for they can be calculated as

$$\langle Z\rangle \equiv \frac{1}{2}\left(\frac{1}{A} + \frac{1}{B}\right) = \frac{1}{2}\frac{2b\beta - a\gamma - c\alpha}{\alpha\gamma - \beta^2};$$

$$\text{var}(Z) \equiv \frac{1}{2}\left(\frac{1}{A^2} + \frac{1}{B^2}\right) \qquad (8.44)$$

$$= \frac{1}{2}\left(\frac{2b\beta - a\gamma - c\alpha}{\alpha\gamma - \beta^2}\right)^2 - \frac{ac - b^2}{\alpha\gamma - \beta^2}.$$

We thus have the interesting conclusion that the essential statistics related to variable Z do not depend but on its coefficients and the values of the curvature parameters characterizing the point of contact. This conclusion can be supplemented with the observation that we have here a purely algebraical basis of selection of the Bessel type wave function related to probability, and supplied by an eigenvalue equation as in the Section 2 above [see equations (2.39) and (2.40)]. The Airy type wave function of Berry and Balazs belongs to this class (Berry & Balazs, 1979)

The Stress by a Statistic

The previous theory can help us secure, from a theoretical point of view, a purely statistical connotation in the curvature space itself. Assume indeed, that 'a', 'b' and 'c' are *some variations* of the curvature parameters α, β and γ respectively, over an ensemble of points locally representing an instantaneous surface inside matter. This instantaneous surface may be a surface proper separating the matter from space, or even an imaginary surface inside matter itself. It turns out that the statistical variable having its values given by equation (8.40) can actually be taken as a variation of the second fundamental form of such a surface, when this variation is controlled only by the variations of its coefficients. Such a situation is particularly important for physical applications. In this case, (8.40) gives the values of a statistical variable — let us call dZ in order to show its 'differential' nature — which has, according to equation (8.44), the expectation

$$\langle dZ \rangle \equiv \frac{1}{2} \frac{A+B}{AB} = \frac{1}{2} \frac{2\beta d\beta - \gamma d\alpha - \alpha d\gamma}{\alpha\gamma - \beta^2} \qquad (8.45)$$

and the variance

$$\langle \left[\Delta(dZ) \right]^2 \rangle \equiv \frac{1}{2} \frac{A^2 + B^2}{A^2 B^2}$$
$$= \frac{1}{2} \left(\frac{2\beta d\beta - \gamma d\alpha - \alpha d\gamma}{\alpha\gamma - \beta^2} \right)^2 - \frac{d\alpha d\gamma - (d\beta)^2}{\alpha\gamma - \beta^2}. \qquad (8.46)$$

These two statistics have a precise geometrical meaning, which may not be obvious by itself to the casual observer. However, if we use them in building another statistic:

$$\langle \left[\Delta(dZ) \right]^2 \rangle - \langle dZ \rangle^2 \equiv \langle dZ^2 \rangle - 2\langle dZ \rangle^2$$
$$= \frac{1}{4} \left(\frac{2\beta d\beta - \gamma d\alpha - \alpha d\gamma}{\alpha\gamma - \beta^2} \right)^2 - \frac{d\alpha d\gamma - (d\beta)^2}{\alpha\gamma - \beta^2} \qquad (8.47)$$

and this statistic has a precise geometrical meaning. First, the right hand side of this formula is the Riemannian metric which can be built by the methods of absolute geometry for the space of the 2×2 matrices, having the curvature matrices with null Gaussian curvature as points of the absolute quadric (Mazilu

& Agop, 2012). This is actually the Klein model of the so-called 'fourth geometry of Poincaré', in the modern views (Duval & Guieu, 1998). Secondly, one can prove that the quadratic form (8.47) is just the Cartan-Killing metric of a homographic action of the 2×2 real symmetric matrices. For, it is, indeed, the quadratic form

$$\frac{1}{4}\left[(\omega^2)^2 - \omega^1\omega^2\right] \tag{8.48}$$

where $\omega^{1,2,3}$ are the 1-forms from equation (8.4), and this quadratic form means $\mathrm{tr}[(\mathbf{b}^{-1}d\mathbf{b})^2]$, where $\mathbf{b}$ is the curvature matrix.

Now, in order to introduce the surface tension in our formalism, we only need to adopt the natural hypothesis that the deformation it induces is to be accounted for by the variation of curvature. The considerations above just show a logical way toward that connection. Specifically, we leave the realm of infinitesimal deformation expressed by an auxiliary vector $|y\rangle$, and adopt its generalization through the apolarity condition (8.11). The vector representing the deformation, is then basically that from equation (8.15). Considering, for the sake of illustration, the small parameter ε, as well as a constant local stretch of the surface metric, included in the variation of the curvature parameters, the metric tensor expressing this deformation is given by equation (8.16) as

$$\begin{pmatrix} 1 + (v^2)^2 & -v_1v_2 \\ -v_1v_2 & 1 + (v^1)^2 \end{pmatrix}. \tag{8.49}$$

The eigenvalues of this matrix are 1 and $1 + \langle v|v\rangle$. The corresponding eigenvectors are

$$|e_1\rangle = \begin{pmatrix} v^1 \\ v^2 \end{pmatrix}; \quad |e_2\rangle = \begin{pmatrix} -v^2 \\ v^1 \end{pmatrix} \tag{8.50}$$

respectively. We choose their components as the convenient differential forms $\phi^{1.2}$ from the definition of surface force in equation (8.55). Accordingly the contact force can be described as *a vector* in the local tangent plane, whose components in the two orthogonal eigendirections of the metric tensor are given by

$$|f\rangle = \begin{pmatrix} d\Omega_1^3 \wedge v^1 + d\Omega_2^3 \wedge v^2 \\ -d\Omega_1^3 \wedge v^2 + d\Omega_2^3 \wedge v^1 \end{pmatrix} \equiv \begin{pmatrix} f_1 \\ f_2 \end{pmatrix}. \tag{8.51}$$

One of these components − let's say the first one − is null according to the definition (8.15): naturally, there is no component of contact force along a direction, if there is no deformation along that direction. As to the other component − in this case the second component − using equation (8.15), it is

$$f_2 = d\Omega_1^3 \wedge [\omega^1 s^1 + (\omega^2/2)s^2] + d\Omega^{2^3} \wedge [(\omega^2/2)s^1 + \omega^3 s^2].$$

146

Assuming now the gauge definition of the variation of curvature, as given by equations (8.2) and (8.3) this contact force becomes

$$f_2 = [(d\lambda - \omega^2/2)\omega^1 - \omega^2\omega^3/2](\Omega_1^3 \wedge s^1)$$

$$+[(d\lambda + \omega^2/2)\omega^2/2 + (\omega^1)^2](\Omega_2^3 \wedge s^1)$$

$$+[(d\lambda - \omega^2/2)\omega^2/2 - (\omega^3)^2](\Omega_1^3 \wedge s^2)$$

$$+[(d\lambda + \omega^2/2)\omega^3 + \omega^1\omega^2/2](\Omega_2^3 \wedge s^2).$$

Recall that we are using here the exterior multiplication in the tangent plane at a point of nucleon surface, not in the space of curvature parameters! With equations (7.8), (8.11) and (8.13) this 2- form shows up as

$$f_2 = \Big\{ d\lambda[(\omega^2/2)(\gamma - \alpha) + (\omega^1 - \omega^3)\beta]$$

$$+[\omega^2/2)^2 - \omega^1\omega^3](\alpha + \gamma)\Big\}(s^1 \wedge s^2).$$

After some lengthy, but otherwise straightforward calculations, based on equation (8.1), we further reduce this force to its final expression:

$$f_2 = (\alpha + \gamma)\Big[d\lambda d\xi + (\omega^2/2)^2 - \omega^1\omega^3\Big](s^1 \wedge s^2) \tag{8.52}$$

where ξ is the logarithm of the ratio of the two curvatures, as defined in equation (8.25).

Just like the classical surface tension, this force contains indeed the mean curvature of the surface in a point. However, unlike that classical expression it also depends on the variance of the increment of the second fundamental form due to a process of 'wrinkling' at the local fragment of surface, a phenomenon expressed by the variations of curvature parameters. If either one of the mean and Gaussian curvatures is not affected by this 'wrinkling', the magnitude of the surface force thus defined can be expressed exclusively by the Riemannian metric of the space of curvature parameters.

The Tensions: Conclusions and Outlook

The classical little cube, serving to prove the existence of tensions, was always a problem in the definition of the stresses in a continuum. The main issue is the fact that this 'gedanken' instrument of mathematics finds itself always in deformation due to the very stresses it is supposed to define. However, using the idea of interpretation here, we can conceive another approach to this issue, avoiding the geometrical precise form, to which we can arrive some other way, involving the deformation itself. Namely the continuum can be interpreted as an ensemble of material points, penetrating each other with or without merging, to use Madelung's expression. A material point inside matter, can be further conceived as a convex body limited in space by an irregular surface in permanent transformation due to interactions with other material points. If it is to have some understanding of the strong and weak forces inside nuclear matter, then we have to describe a very first instance of the interactions between material points, namely the contact. There are two aspects of the geometrical and physical theory of contact: first the a priori choice of the location on the host delimiting surface of the material point taken into consideration, then, secondly, the measure of the contact spot, due to the neighboring material points and their induced forces.

A natural idea about the physical description of the contact on the host surface, is that it can be defined only locally, and only at a certain scale, by the curvature parameters and their changes. This fact has the twofold advantage of being clearly accountable by a mathematical form: on one hand, a deformation of surface which can be expressed by local curvature changes, on the other hand a certain definition of the variation of the local curvature vector. These two mathematical results converge in a logical definition of the contact force as a differential 2-form that generalizes the classical definition of superficial tension.

The space of curvature parameters can be organized as a Riemann space, whose metric, has a precise statistical meaning as the standard deviation of a variation of the second fundamental form with respect to its nominal value, induced by exclusive variation of its coefficients (the curvature parameters). The contact force, as defined here, is proportional to both the nominal mean curvature as in the classical case, but also involves the statistical variance of the second fundamental form, therefore the Riemannian metric of the space of the curvature parameters. This fact can have important consequences in the description of the dynamics of a material point inside matter by a gauge theory. But unlike the classical case, the gauge defined by the variation of local curvature, asks for a proper definition of the instantaneous delimiting surface of a material point. This surface can be conceived as an ensemble of 'elements of contact', whose characteristic contact force already contains a statistical element through the variance of the second fundamental form.

The actual instantaneous delimiting surface of a material point *inside matter* is an issue demanding further elaboration. However, one could say that this elaboration can take advantage of a sound guidance, both from the classical Fresnel theory of the wave surface in the case of light, and from the modern theory of a holographic universe, according to which the interior of a material point should be structured as a hologram [('t Hooft, 1993); (Susskind, 1994)]. In this respect, the actual surface of a material point can even be taken as a fuzzy sphere in the sense of John Madore (Madore, 1991, 1992). This seems to be quite a natural approach, in view of the fact that, with the variation of the curvature parameters as presented in this work, we reach actually in the realm of $\mathbf{sl}(2,\ R)$ Lie algebra. However, our approach will be to consider the basis of a three-dimensional algebra as only a *reference frame* in the sense of Alexey Shchepetilov, for the geometry of curvature parameters (Shchepetilov, 2003). A continuous theory of Frenet-Serret type can then be established [see (Dubois-Violette, Kerner & Madore, 1990) for the general ideas on this issue], which is a genuine holographic theory of material points, and also contains those elements of fuzziness necessary to a stochastic methodology.

Chapter 9. The Nonstationary Description of Matter

If we are anywhere near accepting the stationary Schrödinger equation as a natural instrument of our knowledge unquestionably, the case with the nonstationary Schrödinger equation is however far from being accepted as such. Particularly the idea of free particles in the matter itself seems to be hopeless with any standing, be it classical, wave-mechanical or quantum-mechanical. On the other hand though, the concept of a Madelung fluid certainly pushes our fantasy to the point of accepting that the matter *per se* admits at least a description, if not even a physical structure, by ensembles of free Hertz material particles. It seems, therefore, a duty from our part, to further elucidate the idea of physical *freedom according to the nonstationary Schrödinger equation.*

The Louis de Broglie Moment

The nonstationary Schrödinger equation (2.33) has a solution in the form of the Gaussian (Synge, 1972). For 'n' space dimensions, this Gaussian can be written in the form (Skinner, 2016)

$$\psi(x, t) = \frac{1}{\sqrt{(4\pi\beta t)^n}} \exp\left(i\frac{\mathbf{x}^2}{4\beta t}\right) \qquad (9.1)$$

for positive 't'. In view of the fact that the equation is linear, we can assume the general solution of Schrödinger equation as given in the integral equation form

$$\psi(\mathbf{x}, t) = \frac{1}{\sqrt{(4\pi\beta t)^n}} \int_{-\infty}^{+\infty} u(\mathbf{y}) \exp\left(i\frac{(\mathbf{x} - \mathbf{y})^2}{4\beta t}\right) d^n\mathbf{y} \qquad (9.2)$$

whereby (9.1) plays the role of a kernel. In order to avoid the singularity at t = 0, John Lighton Synge adjusts the time and position with *arbitrary imaginary quantities*, in view of the fact that the equation (2.33) is invariant with respect to translations in time and position and, from physical point of view does not make sense to be singular at the initial moment of time. Based on this, he derives some interesting properties of the wave packets (9.1) that set them in a direct comparison with the de Broglie's wave packets. Quoting:

> My aim has been to present the properties of the [*three-dimensional Gaussian, a.n.*] wave function in a concise and simple way, and to deal with wave packets *concentrated in three dimensions* and not in one dimension only. The velocity **v** appears naturally ... *as the locus of maximum density, rather than as a group velocity*, and *momentum is derived from velocity* ... rather than the other way around. ... the expectation of energy is not simply $(1/2)mv^2$; *there is a suplementary term.* [(Synge, 1972); *our Italics*]

The emphasized conclusions of Synge from this excerpt are of importance: particularly the idea that *the velocity should be somehow related to density*, which seems inconsistent with de Broglie's own relationship between a group of waves and a physical particle. Limiting our discussion to the one dimensional case, rather than to the three dimensional one, we shall concentrate therefore here mostly upon that "locus of maximum density" mentioned by Synge in his conclusion, for this involves directly the idea of density in the sense of Hertz: *the density of material particles.* Naturally then, from this perspective, the equation (9.2) confers to the wave function a property of the Gaussian distribution anticipated by Louis de Broglie himself some three decades before the work of Synge (de Broglie, 1935): pending a proper interpretation of the complex coordinates and time, *it represents the manner in which a field is attached to a physical magnitude.* Let us get into some details of the idea.

Louis de Broglie started from the assumption that the fundamental interactions in the world of material particles involve electromagnetic fields, but he went a little further, methodologically speaking. Namely he moved on to suggest a way in which *a field is applied upon a physical quantity* that can be characterized by a density, and that way seems to us as having universal validity: *it should be true for any kind of field*, in fact it should be the very definition of a field. Specifically, de Broglie used Poisson equation in order to substantiate the idea that the field and matter are generally defined in different positions in space. If V is a field magnitude, defined in a position **x** say, and ρ is the density of an electron, defined in the position **X** say, then, according to classical precepts, the interaction can be expressed by monomials having the following algebraic structure:

$$\rho(\mathbf{X})V(\mathbf{x})\delta(\mathbf{X} - \mathbf{x}) \tag{9.3}$$

where δ is the Dirac symbol. The interaction *per se* is thus expressed by a double space integral over the two space positions

$$\iiint d^3\mathbf{x} \iiint d^3\mathbf{X} \cdot \rho(\mathbf{X})V(\mathbf{x})\delta(\mathbf{X}-\mathbf{x}) = \iiint d^3\mathbf{x} \cdot \rho(\mathbf{x})V(\mathbf{x}) \tag{9.4}$$

so that de Broglie takes note of the fact that in (9.3) and (9.4)...

> ... the factor δ is an ≪ application function ≫ whose role is that of expressing the fact that in each point an *electromagnetic field is applied to the electricity from that point* [(de Broglie, 1935); *our translation and Italics*]

It is not hard to notice then, starting from the very same point of view with Poisson equation, that this ≪ application function ≫ plays also a reciprocal role, so to speak, insofar as it also indicates the way in which, maintaining the de Broglie's phrasing, the "electricity is applied to field". Thus *the field equation* − here the classical Poisson equation − for the field magnitude characterized by the function V, can be written as

$$\nabla^2 V(\mathbf{x}) = \iiint \mathbf{d^3X} \cdot \rho(\mathbf{X})\delta(\mathbf{X} - \mathbf{x}) \qquad (9.5)$$

using the properties of the Dirac symbol. Since in the classical case this leads to the Coulombian potential which is singular in the position of charge, the classical physics was forced to assume here a space extension of the material point, which in turn calls for *a physical structure for the electron*. However, at this point de Broglie takes notice of the fact that

> ... Unfortunately it *does not seem at all possible*, within present-day ideas, *to assign a structure to the electron*: the quantum theories of interaction between electromagnetic field and matter also find an infinite energy for the electron (except, however, the recent very interesting theory of Mr. Born). [(de Broglie, 1935); *our translation and Italics*]

Incidentally, the work of 'Mr. Born' referred to by de Broglie in this excerpt, is the one from 1934, setting the ground for what we know today as the Born-Infeld nonlinear electrodynamics [(Born, 1934); (Born & Infeld, 1934)], of which we shall refer later in this work. Accordingly, de Broglie proposes a way out of impasse, by replacing the symbol δ of Dirac with an isotropic Gaussian:

$$\exp\{-[\mathbf{X} - \mathbf{x})/\boldsymbol{\sigma}]^2\} \qquad (9.6)$$

which reduces to δ in the limit $\sigma \to 0$. In this case, the equation (9.5) takes the form:

$$\nabla^2 V(\mathbf{x}) = \iiint d^3\mathbf{X} \cdot \rho(\mathbf{X}) \exp\{-[\mathbf{X} - \mathbf{x})/\boldsymbol{\sigma}]^2\}. \qquad (9.7)$$

Thus everything happens as if the charge is normally distributed around the point **x**. However, in spite of the fact that it has the dimensions of a length, and even plays the classical part of the radius of a spatially finite electron, the quantity σ ...

... is nevertheless *not a structure parameter*; rather, it is a *parameter of uncertainty* of the position of application of the field upon charge (*or vice versa*). This seems to be in better agreement with the quantum ideas than the *introduction of a genuine radius* [(De Broglie, 1935), *our translation and Italics*]

That 'vice versa' of Louis de Broglie, which we specifically emphasized in this excerpt, expresses an essential point, when combined with his own observation that either the quantum mechanics or the wave mechanics do not support the idea of a "genuine radius" for the electron. It shows that the Gaussian thus introduced is not simply an entirely subjective element: both the application of the field upon charge but, more importantly, *the application of the charge upon field* need to be further documented and physically assessed, on an equal footing. Generally, we can replace here the word 'charge' with 'matter', thus making out of this observation of Louis de Broglie a law. For, by this de Broglie actually introduces the Hertzian element of natural philosophy within the core of theoretical physics. True, only from electrodynamical point of view, but universal as such nonetheless, when referred to matter. Indeed, according to Hertz's ideas, de Broglie's reasoning gives, in fact, explicitly, in the spirit of modern theoretical physics, the way in which a spatially extended particle (an electron, in this case!) *indicates a position in space*: this is 'the position in which *the charge is applied to the field*'. This position is a necessarily random process over a finite space region, with finiteness measured by a *statistical estimator representing the a priori space extension of the electron.* However the normality is only a part of this statistics, and the history of last half of century has brought about facts indicating that the statistics *per se* should be part and parcel of the natural philosophy even from classical point of view.

The second part of our experience with Nottale's SRT, concerns, as we already mentioned, the third order section of nonstationary Schrödinger equation, represented here by equation (1.22). Guided by the soliton interpretation of the Korteveg-de Vries equation, to which (1.22) reduces in the case of one space dimension, we reasoned out that our own equation would represent the dispersion phenomenon within a complex fractal fluid of free particles [(Agop, Păun & Harabagiu, 2008); (Casian-Botez, Agop, Nica, Păun & Munceleanu, 2010)]. Especially the stationary solutions attracted our attention, insofar as they can be represented by special elliptic functions, a feature apparently shared even with the solutions of the classical Laplace equation, defining the potential in free space [(Ouroushev, 1985); (Martinov, Ouroushev & Grigorov, 1991, 1992)]. This would suggest indeed a general underlying statistical interpretation, which was only recently advocated even in theoretical statistics (Letac, 2016). Be it as it may, the interpretation of stationary solutions thus categorized by us does not include the three- dimensional case and, what is more important, those solutions do not make any reference to the nonstationary case. However, historically speaking, there is a nonstationary case, and even related to a *linear third order equation* of the Schrödinger type for that matter, which opens a new direction of research along the idea of a fractal fluid of free particles.

David Vernon Widder took notice of the fact that the *classical Airy function* can be presented as the kernel of a partial differential equation of the first order in time and third order in a space coordinate (Widder, 1979). Namely, the solution of the third order partial differential equation

$$\frac{\partial u}{\partial t} = \frac{\partial^3 u}{\partial x^3} \tag{9.8}$$

can be represented by an integral equation of the form (9.2) for the one-dimensional case:

$$u(x,t) = \int_{-\infty}^{+\infty} K(x-y,t)V(y)dy;$$

$$K(x-y,t) \equiv \frac{1}{(3t)^{1/3}} Ai\left(\frac{y-x}{(3t)^{1/3}}\right) \tag{9.9}$$

with 'V' a function of space variable only, satisfying some convenient conditions imposed by the physics of the problem to be solved, and Ai(...) the Airy function of the first kind. As a ≪ function of application of the field upon matter or of the matter upon field ≫ in the phrasing of Louis de Broglie, the kernel of the integral equation (9.9) satisfies the very same conditions in the classical limit of short times as the Gaussian from equation (9.2) [(Vallée & Soares, 2004), § 4.2, equation (4.16)]:

$$\lim_{t \to 0} K(x-y,t) = \delta(x-y). \tag{9.10}$$

This, in our opinion, gives the strength of a principle to the observation of de Broglie that either the standard deviation in the case of Gaussian, or the period 't' in the case of Airy function, should be taken as 'uncertainty parameters' rather than geometrical quantities in the classical sense. For along this line, the physics brought here a positive case for the uncertainty of the positions in *which the field is applied upon matter*, or vice versa, *the matter is applied upon field*, and this case is referring to the very epitome of the notion of field, the gravitational field.

Airy Moment of Berry and Balazs

Indeed, in this connection we have even a fundamental way to characterize de Broglie's application procedure, closer, we should say, to Hertz's natural philosophy. Michael Berry and Nandor Balazs took note that if in equation (9.2) for the one-dimensional case V(y) *is an Airy function*, then solution $\psi(x,t)$ of the nonstationary Schrödinger equation (2.33) retains this property in a specific and convenient form: *its amplitude is an Airy function* (Berry & Balazs, 1978). Indeed, as an Airy function V(y) is defined by equation:

$$V(y) \equiv Ai(y) = \frac{1}{2\pi} \int_{-\infty}^{+\infty} \exp\left[i\left(\frac{\omega^3}{3} + \omega y\right)\right] d\omega \qquad (9.11)$$

so that equation (9.2) takes the form

$$\psi(x,t) = \frac{1}{2\pi\sqrt{t}} \int_{-\infty}^{+\infty} \exp\left[i\left(\frac{\omega^3}{3} + \omega y + \frac{(x-y)^2}{4\beta t}\right)\right] dy\, d\omega \qquad (9.12)$$

Performing here the integral over 'y', up to a multiplication constant the result is:

$$\psi(x,t) = \frac{1}{2\pi} \int_{-\infty}^{+\infty} \exp\left[i\left(\frac{\omega^3}{3} + \omega x - \beta t \omega^2\right\}\right] d\omega. \qquad (9.13)$$

Now, our final result is obtained using the equation (2.25) from page 10 of (Vallée & Soares, 2004), and it is

$$\psi(x,t) = [Ai(kx - \nu^2 t^2)] \cdot \exp\left[i\nu t\left(kx - \frac{2}{3}\nu^2 t^2\right)\right]; \quad \nu \equiv k^2\beta. \quad (9.14)$$

We only arranged the things in order from a mathematical point of view, within this expression, by the fact that both the argument of the Airy function and the exponent are made explicitly nondimensional. The procedure we followed in doing this is simply by noticing that the exponent of the Gaussian in (9.2) must be non-dimensional, and this property must be preserved by a transformation of the type $x \to kx$, $t \to k^2 t$, which is instrumental in deciding the form of the solution of Schrödinger equation [see (Skinner, 2016), Chapter 4, for the case of heat equation]. Now, if the scale factor 'k' has the dimension of the inverse of a length, i.e. it is either a curvature from a geometrical point of view, or a wave number from a physical point of view, the ratio from exponent in (9.2) is non-dimensional, leading to equation (9.14), whereby ν *has the dimensions of a frequency.* Further on, if we identify this last equation with (2.28), then we have the announced property, for, in that case:

$$A(x,t) = Ai(kx - \nu^2 t^2); \quad \phi(x,t) = \nu t \cdot \left(kx - \frac{2}{3}\nu^2 t^2\right). \qquad (9.15)$$

According to this identification, the potential defined by equation (2.35) is

$$V(x,t) \equiv \frac{\beta}{A(x,t)} \frac{\partial^2 A(x,t)}{\partial x^2} \quad \therefore \quad V(x,t) = \nu \cdot (kx - \nu^2 t^2) \qquad (9.16)$$

155

and classically represents a constant force. Thus, the nonstationary Schrödinger equation, in its 'universal' instance given by equation (M.24), actually produces results related to a *uniformly accelerated motion* (the arguments of the Airy function and that of the potential). This conclusion is valid indeed, provided we take for granted the other one of de Broglie's ideas, referring to the ≪ undulatory phenomenon called classical material point ≫.

Indeed, the original physical interpretation of the result just presented here in its essential lines mathematically speaking [see (Berry & Balazs, 1979) for details], stands upon the very nature of the invention of the Airy function: the behavior of light in the proximity of caustic (Airy, 1838, 1848). We thus have the interesting conclusion that there is a probability density given by the square of the Airy function, i.e. by the square of the amplitude of the wave function, indeed. We might say that this is the very spirit of Louis de Broglie's idea (de Broglie, 1927). However, as this probability density is not integrable over the whole *a priori* real range [see (Valle & Soares, 2004), § 3.5.1, p. 50], the wave packet cannot have a localizable center in the sense of de Broglie, in order to represent a classical material point. The conclusion expressed by Berry and Balazs is that equation (9.14) represents in fact *an ensemble of particles* moving uniformly in straight lines, but with different velocities. The argument of Airy function actually represents a caustic indeed, but in the phase space: it is the *envelope of the ensemble of straight lines representing the corresponding uniform motions.*

A conclusion apparently contradictory to this one is offered by Daniel Greenberger, who gets it by appealing to the *equivalence principle* (Greenberger, 1980): the Airy wave packet is not spreading because *it represents a particle in an enclosure* analogous to Einstein elevator falling in a gravitational field, and thus the gravitational force is suppressed. For the details of analysis of the nonstationary Schrödinger equation at this juncture, one can follow the works indicated by Greenberger himself, especially (Rosen, 1972). This last work of Gerald Rosen tackles the details of the circumstance that Schrödinger equation involves, over the Galilei group, another group of SL(2,R) type in two variables with three parameters. One can thus prove [see (Olver, 1998), Example 3.17, p. 208] that the solution of the Schrödinger equation admits indeed an expression by Airy functions up to a phase factor. However, we are set here on proving that there is no contradiction between the two interpretations: that of Berry-Balazs and that of Greenberger. In fact, by pushing the Berry-Balasz geometry of the phase plane a little further, we can have actually *a physical model of an Einstein elevator*, to be described in detail as follows.

Let us notice first that the Galilei relation, originally giving the speed of a free falling body in a constant gravitational field:

$$v^2 = v_0^2 + 2g(x - x_0) \qquad (9.17)$$

represents perhaps the first relation deserving the name of a *trajectory in the phase plane*, and it is referring to a uniformly accelerated material point. Here 'v' is the speed at position 'x', where 'x' is the height with respect to Earth, up to a sign of course, and 'g' is the gravitational acceleration assumed a constant. The phase plane is then coordinated by the pairs (x,v), reprezenting the position and its corresponding velocity, and the trajectory in this plane is represented by a parabola, as in (9.17). Let us state it again: this phase plane is classic indeed, however not with reference to the uniform motion, but to the uniformly accelerated motion. Physically speaking the relation (9.17) is indeed correct, being dimensionally homogeneous: both terms from the right hand side have dimensions of square velocity, just as the left hand side of the equation. Again, geometrically it represents, in an implicit form, a parabola having the explicit parametric representation given by equations:

$$x(t) = x_0 + v_0 t + (1/2)gt^2; \quad v(t) = v_0 + gt \qquad (9.18)$$

where 't' is the Galileian time.

Now, in this context, the Berry-Balazs argument regarding the solution (9.14) of the nonstationary Schrödinger equation comes down to the idea that the parabola (9.17) is geometrically *an evolute*, inasmuch as a caustic is the envelope of a family of straight lines in the phase plane, representing *unlimited* uniform motions. These, however, are not representative for real motions of the classical material points. This is actually the physical ground of the conclusion that the Airy packet does not represent a material point, but an ensemble of classical material points, as a true interpretation in the sense of Charles Galton Darwin would ask. Now, we preserve here the classical character of the treatment, and even the conclusions, but push the geometry a little further, by presenting the Galilei equation (9.17) as *an involute*, namely the evolvent of a cubic parabola, having the typical form given by the cubic representing the phase of the Airy packet (9.14). Therefore, the ensemble necessary to a physical interpretation referred to by Berry and Balazs is not quite an ensemble of free classical material points, but an *ensemble of Hertz material particles*, having straight uniform motions indeed. However, these motions are confined to segments measured, for instance, by the radii of curvature of the cubic parabola representing the phase in (9.14). This would then explain why the potential in (9.16) is linear in coordinate: it represents actually the whole energy of a family of Hertz *material particles in free fall* in a constant gravitational field [(Synge, 1972); *see his equation* (13)]. This seem to be the most comprehensive physical interpretation of the third point of what we have called the 'Berry moment' of human knowledge – the Airy moment.

In order to complete the technical part of this whole story, one can start with the observation that the parametric equations of the evolvent of parabola (9.18) is a plane cubic. Indeed, the parametric equations of an evolvent are, in their general form given as [(Gheorghiu, 1964), §1.9, equation (3), p. 55, and example 2, p. 58]:

$$X(t) = x(t) - \dot{y}\frac{\dot{x}^2 + \dot{y}^2}{\dot{x}\ddot{y} - \ddot{t}\ddot{x}}; \quad Y(t) = y(t) + \dot{x}\frac{\dot{x}^2 + \dot{y}^2}{\dot{x}\ddot{y} - \ddot{t}\ddot{x}}. \tag{9.19}$$

To wit, here x(t) is the position on parabola in the phase plane, and y(t) is the second coordinate in the phase plane, given by the equation

$$y(t) = y_0 + (g\tau)t; \quad y_0 \equiv v_0\tau. \tag{9.20}$$

Further on, the dot over the symbol means time derivative, as usual, and τ is a constant having the physical dimensions of a time period, necessary in order to establish to the coordinate 'y' the character of a length to be used in geometric calculations. For, in a geometry of the phase plane, the coordinates must obviously have the same physical dimensions, and we chose to work with lengths. After a succession of routine calculations, the formulas (9.19) give the final result

$$X(t) = x_0 + g\tau^2 - \frac{v_0^2}{2g} + \frac{3}{2g}(v_0 + gt)^2; \quad Y(t) = -\frac{(v_0 + gt)^3}{g^2\tau} \tag{9.21}$$

which represents indeed a cubic parabola. In implicit form its equation is:

$$8(X - x_0 - g\tau^2 + v_0^2/2g)^3 = 27g\tau^2 Y^2. \tag{9.22}$$

Therefore, at least qualitatively speaking at this moment, the classical type arguments of Berry and Balazs do have indeed the physical reason which their authors invoke, but only under the condition that the second coordinate in the phase plane is to be taken as the phase of the wave function obtained as a solution of the nonstationary Schrödinger equation, in the initial condition given by an Airy packet.

Let us now reverse the argument, taking the undulatory point of view as a basis. In the present circumstance this point of view has the same rank within our knowledge with the classical dynamical point of view, and consequently it should be just as important. Indeed, we know for sure that the Airy packet (9.14) is a solution of the nonstationary Schrödinger equation for the free particle (Berry & Balazs, 1979). This knowledge is, mathematically speaking, just as sure as is the theory leading to the Galilei equations (9.18). Therefore the signal (9.14) can be taken as a general signal, to which we have however the obligation to find a physical interpretation. Extending the observations of Berry and Balazs by taking into considerations just motions of material particles which extend over finite geometrical ranges, it would then be necessary to find the involute of the cubic parabola represented by parametric equations:

$$X(t) = x - \frac{1}{2}at^2; \quad Y(t) = \nu t\left(x - \frac{1}{3}at^2\right) \qquad (9.23)$$

where the quantity

$$a \equiv \frac{2\nu^2}{k} \qquad (9.24)$$

represents physically an acceleration. The parametric equations of the involute of this cubic parabola are taken according to a general formula of calculation of an involute [(Guggenheimer, 1977); the equation (37), p. 39]:

$$\xi(t) = X(t) - s(t)\frac{\dot{X}(t)}{\sqrt{\dot{X}^2 + \dot{Y}^2}}; \quad \eta(t) = Y(t) - s(t)\frac{\dot{Y}(t)}{\sqrt{\dot{X}^2 + \dot{Y}^2}}. \qquad (9.25)$$

Here s(t) is *the arclength of the portion of the cubic parabola* whose involute we need to find. Let us calculate these coordinates. We shall carry out the calculations exclusively for the case x — 0 in equation (9.23): the general case, $x \neq 0$, is entirely analogous, but with more involved calculations, and the results of principle are exactly the same as for this particular case. Thus, in the interest of clarity in expounding of the very point of view, it becomes almost necessary, we should say, to simplify the calculations, by limiting our considerations to the case x = 0. Therefore, for x = 0 in (9.23), we have:

$$\dot{X}(t) = -at; \quad \dot{Y}(t) = -a\nu t^2 \qquad (9.26)$$

and with these we have directly the following expression for the arclength:

$$s(t) \equiv \int_0^t du\sqrt{\dot{X}^2(u) + \dot{Y}^2(u)} = \frac{a}{3\nu^2}\left[(1 + \nu^2 t^2)^{3/2} - 1\right]. \qquad (9.27)$$

Using now the results (9.26) and (9.27) in (9.25) we have finally:

$$\xi(t) = \frac{a}{3\nu^2} - \frac{1}{6}at^2 - \frac{a}{3\nu^2}\frac{1}{\sqrt{1 + \nu^2 t^2}};$$

$$\eta(t) = \frac{a}{3\nu}t - \frac{a}{3\nu^2}\frac{\nu t}{\sqrt{1 + \nu^2 t^2}}. \qquad (9.28)$$

Obviously, we do not have here a Galilean parabola or, better, we do not have it always. Indeed it cannot be obtained quite unconditionally insofar as the value of parameter 'a' is concerned, but only in cases where the frequency ν is sufficiently high in order to be possible to neglect its inverse square by comparison with the inverse itself in (9.28). In these cases we have indeed the parabola:

$$\xi(t) = -\frac{1}{6}at^2; \quad \eta(t) = \frac{a}{3\nu}t. \tag{9.29}$$

This parabola represents an accelerated motion in the phase plane, characteristic, for instance, to a free fall with the acceleration $g = a/3$ and initial velocity $v_0 = a/(3\nu)$. These parameters define a time period $v_0/g = 1/\nu$ which should be small enough in order to satisfy the required conditions.

Therefore, as an involute, the curve obtained from the solution of nonstationary Schrödinger equation by a 'Berry-Balazs method' amended in the way just shown, is not quite the parabola of free fall, but a cubic curve, having however a well defined connection with that parabola. Namely, the equation (9.28) represents a *curve parallel with the free fall parabola*, with the parallelism defined in a geometrically precise sense. For, in the present context, the free fall parabola is not the one given by equation (9.29), but the curve having the parametric equations:

$$\xi(t) = \frac{a}{3\nu^2} - \frac{1}{6}at^2; \quad \eta(t) = \frac{a}{3\nu}t. \tag{9.30}$$

The unit normal to this curve has the parametric components

$$\frac{1}{\sqrt{1 + \nu^2t^2}}; \quad \frac{\nu t}{\sqrt{1 + \nu^2t^2}} \tag{9.31}$$

so that the curve (9.28) is the locus of the positions in the phase plane located on the normals of (9.30) at the constant distance $d = -a/(3\nu^2)$.

Thus, let us just assume that we have to start afresh from the nonstationary Schrödinger equation, taken this time as *an universal instrument of our knowledge*, in order to construct a fluid dynamics of the Hertz material particles, for interpreting the wave mechanics. Having initially an Airy packet in the construction of the wave function (9.2), we describe not an *ensemble of uniform motions*, but an *ensemble of motions having a definite relationship with a uniformly accelerated motion*. Represented in the phase plane in the geometric manner of Berry and Balazs, but in action-angle variables, this ensemble is characterized by positions along the normals to the trajectory representing the uniformly accelerated motion. The locus of these positions is a trajectory in the phase plane parallel to the parabola of uniformly accelerated motion, with the parallelism defined by the distance along normals. In fact, writing the difference between the two positions in the form:

$$
\begin{aligned}
\frac{a}{3\nu^2} - \frac{1}{6}at^2 - \xi(t) &= \frac{a}{3\nu^2}\frac{1}{\sqrt{1 + \nu^2t^2}}; \\
\frac{a}{3\nu}t - \eta(t) &= \frac{a}{3\nu^2}\frac{\nu t}{\sqrt{1 + \nu^2t^2}}
\end{aligned}
\tag{9.32}
$$

we even have a physical meaning of the classical Galilean time of the uniformly accelerated motion, or in fact of the free fall: *it is the time of some known geodesic motions.* Indeed, if we designate $\nu t = \tan\theta$, then these equations become:

$$\frac{a}{3\nu^2}\tan\theta - \eta(\theta) = \frac{a}{3\nu^2}\cos\theta;$$

$$\frac{a}{3\nu^2} - \frac{1}{2}\frac{a}{3\nu^2}\tan^2\theta - \xi(\theta) = \frac{a}{3\nu^2}\sin\theta. \tag{9.33}$$

This shows that the difference between the motion of such a material particle and the uniformly accelerated motion, is represented by a harmonic oscillator, having the phase $\tan^{-1}(\nu t)$.

In order to better grasp the physical meaning of this situation, let us summarize the way of reasoning we followed up to this point: the nonstationary Schrödinger equation (2.33) describes an ensemble of Hertz material particles. Judging by the character of this partial differential equation, these are free particles, qualified therefore to describe the structure of a complex fractal fluid, as SRT requires. If the starting conditions in the constructing the solution of the Schrödinger equation are given by an ensemble described by a density which is the square of an Airy function, then this solution describes an ensemble of Hertz particles characterized *in the phase plane* by a constant Euclidean distance from the positions of a uniformly accelerated motion. Therefore, when we say *Berry-Balazs free particle*, this is by no means a *Galilei free particle*. Fact is that the equation (9.33) is susceptible indeed, to a purely ensemble interpretation, from a statistical theoretical point of view.

Cosmological Moment of Berry and Klein

The documentation of this last statement takes us, again, to the third achievement listed by us under what we designated as the *Berry moment* of modern positive knowledge. In connection with the idea of the free fall of an Einstein elevator, the need of such a documentation goes, as we have already mentioned, as deep in the history of human knowledge as the Newton's invention of the forces responsible for the gravitational action. First, however, let us see what is meant here by an Einstein elevator: in the light of the conclusions right above, *we can imagine an enclosure*, just like the Wien-Lummer enclosure, *containing only matter and no space.* This may be, for instance, an elementary particle falling in the gravitational field of Earth, or revolving in a Kepler problem with extended bodies. Then, according to Berry-Balazs theory above, there is just a position inside matter, which at a certain moment of time has an instantaneous well defined acceleration [for details on this issue see (Mazilu & Porumbreanu, 2018)]. With respect to *that spacetime event*, there exist in the enclosure an ensemble of material particles having constant Euclidean distance, which manifest themselves as harmonic oscillators at a certain time scale. This point of view in the description of such a structure is entirely justified, at least speculatively speaking, in the case of nucleus of the planetary model, by the results of the so-called *focal regularization* (Burdet, 1968, 1969). Fact is that the manner of procedure can be documented even starting with Galilean studies of motion, by statistical methods based on equations (2.45) and (2.46) from the Section 2 here [see, for details, (Mazilu & Porumbreanu, 2018)]. Thus, it can be shown that the only forces compatible with this situation are the Newtonian forces with magnitude inversely proportional with the square of distance between the Hertz material particles (Berry & Klein, 1984).

In the present context, the great merit of the work of Berry and Klein just cited here, appears to be the use of a gauging argument of the type that led to the Wien displacement law, and this is what really makes the *Wien-Lummer enclosure* and *Einstein elevator* two different expressions of the same concept: *the physical reference frame.* However, as observed before, insofar as the two instances of physical reference frame characterize two different productions of the positive knowledge, they also represent mathematically two completely different transitions. To wit, in view of the Berry-Klein theory, the Einstein elevator can be taken as carrying the burden of transition between microcosmos and the world of our daily experience, while the Wien-Lummer enclosure carries the burden of transition between the world of daily experience and universe at large. This physical conclusion is based upon mathematical observation that the Newtonian forces are the mark of the world of daily experience – the quotidian world. They are justified even by Newton himself by collision forces which, in the phrasing of Nicholas Georgescu- Roegen, acting over *infrafinite time intervals*, are the clear expression of the microcosmos. On the other hand, the Wien-Lummer enclosure is justified by the necessity of indefinite extension of our daily experience to the whole universe. Therefore, mathematically, it should represent a *transition between finite and transfinite*. As the measurement of light have also decided the peculiarities of the physical structure of microcosmos, the Wien-Lummer enclosure carries a mark of universality, also manifest by the fact that it allows thermodynamic judgments, referring to adiabaticity for instance. It seems, however, that the Berry-Klein theory of scaling the forces, *in case it can be applied to a Wien-Lummer enclosure*, acts as a *universal gauging procedure* which, in an undertaking of Georgescu-Roegen type, should be able to make us decide what is infrafinite, finite or transfinite. Let us, therefore, get into some details of that theory.

In the present circumstance, the essential point of Berry's and Klein's work can be summarized as follows: consider a Hertz material particle in a field of forces represented by a potential $V(\mathbf{r})$, with its motion characterized by a Hamiltonian of the form:

$$H(\mathbf{r}, \mathbf{p}; \ell) = \frac{\mathbf{p}^2}{2m} + \alpha(\ell)V(\mathbf{r}/\ell). \qquad (9.34)$$

Here the length ℓ itself is a function of the time of motion, and epitomizes the idea of isotropic expansion, the way it was described by us for the case of blackbody radiation. To wit, one assumes that every quantity which is length in a certain enclosure, no matter of its further physical attribute, is affected in expansion by the factor ℓ depending on the time of motion which, according to classical precepts is a unique time in a Newtonian universe. The invariance of the field of forces represented in the Hamiltonian (9.34) by the potential $V(\mathbf{r})$, from which the force itself is derived via the classical recipe $\mathbf{f}(\mathbf{r}) \equiv \nabla_{\mathbf{r}}V(\mathbf{r})$, can here be expressed in the form:

$$\alpha(\ell)\mathbf{f}(\mathbf{r}/\ell) = \mathbf{f}(\mathbf{r}). \qquad (9.35)$$

For the inverse square forces, this invariance condition comes down to a more precise definition of the function $\alpha(\ell)$:

$$\ell^2 \cdot \alpha(\ell) = 1. \qquad (9.36)$$

In words: if the function $\alpha(\ell)$ is inversely proportional with the square of the gauge length, *the Newtonian force has the property defining the Wien displacement law*, i.e. it is invariant with respect to the expansion, no matter if this is adiabatic or not.

Let us analyze the motion from the point of view of the classical dynamics, following the work of Berry and Klein. Hamilton's equations: $\dot{\mathbf{r}} = \nabla_p H, \dot{\mathbf{p}} = \nabla_{\mathbf{r}}H$, come down to the classical equation of motion:

$$m\ddot{\mathbf{r}} + \alpha(\ell)\nabla_{\mathbf{r}}V(\mathbf{r}/\ell) = \mathbf{0}. \qquad (9.37)$$

In a general continuity parameter $\tau(t)$, and for the positions in a Euclidean reference frame redefined by $\mathbf{x} \equiv \mathbf{r}/\ell$, the equation of motion assumes the form

$$\frac{m}{\alpha}\left[(\ell\dot{\tau})^2\mathbf{x}'' + \frac{d(\ell^2\dot{\tau})}{dt}\mathbf{x}' + (\ell\ddot{\ell})\mathbf{x}\right] + \nabla_{\mathbf{x}}V(\mathbf{x}) = 0 \qquad (9.38)$$

where the accent means the derivative upon time τ. Under condition (9.35), this becomes

$$m\left[(\ell^2\dot{\tau})^2\mathbf{x}'' + \ell^2\frac{d(\ell^2\dot{\tau})}{dt}\mathbf{x}' + (\ell\ddot{\ell})\mathbf{x}\right] + \nabla_{\mathbf{x}}V(\mathbf{x}) = 0. \qquad (9.39)$$

163

Now, we redefine the time itself by $t' = \ell^2$, so that this equation of motion simplifies to one with *no dissipative* forces:

$$m\mathbf{x}'' + m(\ell^3 \ddot{\ell})\mathbf{x} + \nabla_{\mathbf{x}} V(\mathbf{x}) = \mathbf{0}. \tag{9.40}$$

Finally, if the *gauge length is determined by the inertial mass* of the material particle such that:

$$m(\ell^3 \ddot{\ell}) = k \tag{9.41}$$

where 'k' is an appropriate physical constant, then the continuity in the τ parameter is defined via a dynamics, and thus it deserves indeed the name of time, forasmuch as it is referring to a *dynamics in a conservative force field*:

$$m\mathbf{x}'' + \nabla_{\mathbf{x}}[V(\mathbf{x}) + (k/2)\mathbf{x}^2] = \mathbf{0}. \tag{9.42}$$

This is a situation as the one just presented beforehand, in connection with the Berry-Balazs theory of Airy wave packets: the position of free fall in matter is accompanied by harmonic oscillators. It should, therefore, be noticed that the conservative forces, initially represented by the potential $V(\mathbf{x})$, are now amended by elastic forces in the gauged coordinates.

The equation (9.41) represents the essence of the Berry-Klein gauging procedure, and allows a precise characterization of such a procedure. Indeed, the solution of differential equation (9.41) has a well-known history mostly concentrated in the second part of the last century. In short, for what we have to show here the result can be conveniently summarized as follows (Eliezer & Gray, 1976): if 'u' and 'v' are two independent solutions of the same ordinary second-order differential equation

$$mu'' + ku = 0, \quad mv'' + kv = 0,$$

then ℓ calculated via quadratic form $\ell^2 \equiv au^2 + 2buv + cv^2$, is a solution of (9.41), provided

$$ac - b^2 = k/m. \tag{9.43}$$

Therefore, the continuity parameter τ is a time indeed, defined by a dynamics corresponding to a motion in the conservative field $V(\mathbf{x})$, over which a harmonic oscillatory motion is superimposed. An intuitive explanation of this situation should be helpful at this moment of our argument.

To the extent to which it can be geometrically explained, the usual eye view can obviously be relegated to the idea of a pencil of geometrical directions in the form of a cone. We presented this idea by following it in the details of current context (Mazilu & Porumbreanu, 2018), but there are exquisite classical presentations [see (Coddington, 1829), especially pp. 1 - 4, for a summary of the geometrical definitions involved in the classical geometrical optics]. Fact is that this purely geometrical image of a physical light ray fits in detail the present day concept. The continuous tracking of the motion of a classical material point — therefore of a Hertz material particle — from a certain position, is always realized only by a geometrical projection of that material point on a surface. The position of surface remains however, undecided, as long as we don't know the distance where the material particle is located with respect to the position of observation. Let us assume though, that by certain physical means we are able to know that distance, and it corresponds to a position given by the vector **r**, which defines a family of Cartesian reference frames having common origin in the point of observation. Even by this, the position as such remains undecided: there are at least a double infinity of Cartesian reference frames describing it. We can choose from this double infinity one of the reference frames, for instance by a signal propagation procedure.

This procedure can certainly be used — in fact it is almost exclusively the procedure of choice in physics — if the physical process that we have at our disposal for observation is the propagation of light, or the propagation of any other perturbation in fact, which is describable by an equation of propagation. Only, we have to be careful, because such a procedure offers, as a rule, a position in space which is largely a matter of some assumptions. Usually such assumptions are referring to the propagation itself, to the source of perturbation and such, but its main point may have been misplaced, if we are to take into consideration the wave mechanics, as we shall see shortly by the way of a significant example. In this case we need to devise a procedure for extracting known coordinates — the coordinates of the observation point — and this procedure should actually include the meaning of coordinates, the meaning of their origin and their connection with the space filled by matter. In order to make these issues clear, we have to cast a critical eye over one of the most important problems occuring along the way of relating the science with the social practice, namely the *problem of transport*, which is intrinsically related to the general concept of continuity.

Chapter 10. The Idea of Continuity in Fluid Dynamics

Nottale's idea contains an implicit concept, according to which the continuity must be adopted by an *adaptation*: at a certain scale, *the continuity has to be described in connection with the continuity with respect to a previous scale.* This is actually the case of the classical Ehrenfest theorem. The problem here is the general description of the *transition between scales of continuity*, which is mandatory in deciding the categories of mathematical order infrafinite-finite-transfinite, according to the ideas of Nicholas Georgescu-Roegen. It seems that the Berry-Klein theory opens a way of dealing with the problem of this decision along the concept of interpretation, but there is a drawback. Namely, from physical point of view, the theory of ensembles, as it is needed for accomplishing an interpretation, involves quite a few intricate concepts: *transport theory*, as a mean of transiting between scales, the *deformation theory* and an associated 'general rotation theory', for maintaining the description within the same scale, and a host of other notions related to these. In order to properly understand the physics' point of view in the concept of interpretation, it is therefore necessary to appeal indeed to the transport theory in fluids, because only this way we can connect the concept of *Madelung fluid* with the Newton's idea of *matter filling a space*.

The Mass Transport in a Volume Element

A first step in an adaptation process of the kind envisioned by Laurent Nottale, would be a continuity equation telling us just how continuous is a system in space. In the cases where such a system does not contain space in its structure, the Newtonian-type continuity is to be taken into consideration. In usual differential form, the content of such an equation can be expressed in words, by simply saying that the mass element is preserved by transport. Thus, if the transport is done in a time sequence t that sets in order the states of matter contained in a volume element, then the equation of continuity can be written as

$$\rho dV|_{t=0} = \rho dV|_t. \tag{10.1}$$

Here the initial time moment 0 is arbitrarily chosen, and 't' is a time moment reckoned with respect to this initial moment. This equation does not mean much if we do not take into account the specific transport, viz. *that transport which sets the time as a sequence*, thus involving the Lie derivative based on the vectorial field along which the transport is accomplished.

Consequently, the first in order of things here, comes the fact that the volume element must be oriented, in order to be properly considered for the definition of the time sequence: it becomes a differential 3-form, while the density becomes a third order tensor. From the entries of this tensor, the Newtonian definition of the mass element samples out only a totally antisymmetric part, i.e. a third order tensor skew-symmetrical in every pair of its three indices:

$$\rho dV \equiv \rho_{klm}(t, \mathbf{x})(dx^k \wedge dx^l \wedge dx^m). \tag{10.2}$$

Here a summation over repeated indices is understood, as usual. The scalar ρ in the left hand side of the equality must be then a certain invariant of the tensor $\boldsymbol{\rho}$. This is obtained from the right hand side of (10.2) if we decompose the sum into appropriate partial sums, defined according to the orientation of the volume element 3-form:

$$\rho \equiv \frac{1}{2}[\rho_{(123)} - \rho_{(312)}]; \quad dV \equiv dx^1 \wedge dx^2 \wedge dx^3. \tag{10.3}$$

The symbol $\rho_{(klm)}$ represents the arithmetic mean of the elements of tensor $\boldsymbol{\rho}$ over the three different permutations of the indices, having the same parity.

A digression seems in order here: according to the orientation of the volume element there are two kinds of scalar densities of matter, represented in equation (10.3) by the two arithmetic means of the entries of skew-symmetric part of the density tensor. It is therefore worth keeping in mind that the density is, in general, a third order tensor, from which the idea of an equation of continuity according to Newton's definition samples out, as we said, only a skew-symmetric part. If the Louis de Broglie's theory of the *proportionality of density with the square of field amplitude* is true, then, according to equation (10.3), there should be, correspondingly, *two kinds of fields providing amplitudes* for this equivalence of the continuum with a field. As long as the D'Alembert equation is involved, the things are classically illustrated by the electromagnetic field in vacuum. Indeed, in this case, the field intensities are both solutions of the homogeneous D'Alembert equation [in a Lorenz gauge freedom, of course (Lorenz, 1867)]. Therefore one orientation of the volume element would correspond to electric field, $\mathbf{E}$ say, while the other one would correspond to magnetic field $\mathbf{B}$, so that a de Broglie-type theory offers here a density of the form

$$\rho \equiv \frac{1}{2}(\mathbf{E}^2 - \mathbf{B}^2)(dx^1 \wedge dx^2 \wedge dx^3) \tag{10.4}$$

where the squares are understood up to a physical factor. The two densities can be represented indeed by time sequence means (Gibbs, 1883). The case of an electromagnetic field in matter elucidates things even further here, inasmuch as it shows that the physical constant, necessary for the equivalence between matter and field, splits into two different constants, one for electric field, viz. *electric permittivity*, and one for the magnetic field, viz. *magnetic permeability*. As it turns out in this case *only one of these constants can represent a density*, and the experimental data then cannot but only help decide which one of them (Fessenden, 1900).

Now, the equation (10.1) itself is the finite time form corresponding to a differential conservation law, which can be given by the natural assumption that the evolution of the mass element happens in space along a vector, $\mathbf{X}$ say. The conservation (10.1) taken over an infinitesimal time interval, is then expressed by the vanishing of the Lie derivative along this vector, which by exterior differential equation (6.11), comes down to:

$$d \wedge i_x(\rho_{(klm)}dx^k \wedge dx^l \wedge dx^m) = 0. \qquad (10.5)$$

Here $i_x(\rho dV)$ signifies the projection of the differential form taken as argument along the vector $\mathbf{X}$. Therefore, the evolution of the differential 3-form (10.2) in space *must be a Hamiltonian evolution* imposed by the Poincaré classical lemma: there is always a differential form in terms of which the differential form $i_x\rho dV$ from equation (10.5) is its exact differential. As we have shown before − see equations (6.10) to (6.19) − this property further triggers some other properties imposed by that Hamiltonian evolution upon the vector field along which the mass is carried, in the following manner (Dumachev, 2009, 2010, 2011).

Assume first that $\mathbf{X}$ is a vector proper, having the meaning − which seems appropriate in this context − of a 'velocity' that describes the connection of the transport process with a certain time sequence indexing the states of the transported structure. This velocity field, $\mathbf{w}$ say, must be tangent to a current line along which the transport is accomplished:

$$\mathbf{X} \equiv \mathbf{w}; \quad \mathbf{w} = w^k \frac{\partial}{\partial x^k}. \qquad (10.6)$$

Then by the projection operation from equation (10.5) we can infer that the mass is 'laid-down' in the volume element on a 'support' offered by the curve locally represented by the vector $\mathbf{w}$. Then it is transported along this curve in the form of a 'mass flux':

$$i_\mathbf{x}(\rho_{klm}dx^k \wedge dx^l \wedge dx^m) = (\rho_{klm}w^k)(dx^l \wedge dx^m). \qquad (10.7)$$

168

In this case, the equation (10.5) shows, via Poincaré lemma, that there is a differential 1-form, $v \equiv v_k dx^k$ say, 'v' − from velocity! − such that

$$(\rho_{klm} w^k)(dx^l \wedge dx^m) = d \wedge v. \tag{10.8}$$

Therefore this 'mass flux' can be physically classified as *a vortex*, and mathematically represented by the curl operation on the vector defining the form 'v':

$$\rho_{klm} w^k = \omega_{lm} \equiv \frac{\partial v_l}{\partial x^m} - \frac{\partial v_m}{\partial x^l}. \tag{10.9}$$

Assuming, in the second place, that the mass is 'laid-down' on a support offered this time by a bivector **w**:

$$\mathbf{X} \equiv \mathbf{w}; \quad \mathbf{w} = w^{ij} \frac{\partial}{\partial x^i} \wedge \frac{\partial}{\partial x^j}, \tag{10.10}$$

instead of (10.7) we will have a differential 1-form

$$i_{\mathbf{x}}(\rho_{(klm)} dx^k \wedge dx^l \wedge dx^m) = (\rho_{klm} w^{lm}) dx^k. \tag{10.11}$$

In words: if the mass is laid-down in the volume element on a surface locally characterized by the bivector **w**, it is then transported by this surface in the form of elementary work of a certain vector. The equation (10.5) then shows that there is a potential function 'ϕ', with respect to which this vector can be expressed as a gradient, i.e.

$$\rho_{klm} w^{lm} = \frac{\partial \phi}{\partial x^k}. \tag{10.12}$$

This potential represents a function whose *level surfaces* in space are surfaces upon which the mass is laid-down in the volume element, and thus it is transported throughout the space. Finally, if the mass is 'laid-down' in the volume element on a trivector **w**:

$$\mathbf{X} \equiv \mathbf{w}; \quad \mathbf{w} = w^{ijk} \frac{\partial}{\partial x^i} \wedge \frac{\partial}{\partial x^j} \wedge \frac{\partial}{\partial x^k}, \tag{10.13}$$

then it is transported as a scalar:

$$\rho = \rho_{ijk} w^{ijk} \tag{10.14}$$

so that the transport maintains this scalar constant during transport. Thus, if the ambient manifold describing the possible phase space is three-dimensional, there are just three possibilities in which the mass can be carried in this phase space: on a line, on a surface and in points given by positions in space.

The previous theory can be understood not quite so much by conservation laws, as much as it can by the way in which a continuum is filled by continuity at a differential level. The equations (10.9), (10.12) and (10.14), can be taken as conservation laws indeed, but they tell us something more: such a law depends on the kind of continuity we conceive to be preserved by transport. In other words, these conservation laws offer in fact the necessary expressions of the magnitudes from the left hand side, as provided by the quantities from the right hand side. These are: a vortex, a gradient and a scalar respectively. Therefore – by the way of anticipating some further results here – if the mass itself is a carrier of some physical quantities like charge, momentum and such like, and the mass element is to be preserved during transport, then it cannot carry those physical properties but on a surface, on a curve or in a point of the volume in which the matter is accounted for. The conservation of the mass element thus forces *the shape of the carriers through which the mass itself carries some other physical quantities.*

Transport Theorem in Finite Volume

The previous transport theory characterizes the infinitesimal transport in time and space: *infrafinite* in the phrasing of Georgescu-Roegen. The things change significantly if the transport occurs in a *finite volume*, usually termed as *control volume* in engineering terminology, for obvious reasons: from such a volume we extract, or in it we inject, physical properties of industrial and economical interest, like heat, momentum, charge, etc. And such a control volume cannot be infinitesimal in engineering practice, which is in fact the social expression of our daily experience. One can claim the control over an elementary volume, however not a first-hand control, but by the intermediary of some mechanical laws, well established indeed from mathematical point of view. Thus, for instance, according to previous observations, *if the evolution is Hamiltonian, one can claim control over the elementary volume*, which thus becomes itself a control volume.

Restricting, therefore, for the moment being, our considerations to the practical case of transport in a finite control volume, the mass of a certain continuous system contained in this volume can be calculated by an integral:

$$m_{System} = \iiint_{Volume} \rho \, dV. \qquad (10.15)$$

Now, a physical system can transport something only if it moves, and during this motion it changes both the configuration and the structure. In the case of continuous systems this motion can be characterized by the fact that between system and volume there is some discrepancy. This discrepancy means that if at a given moment of time the system occupies a certain volume, at a later time it occupies another volume. Physically, the system may be the same, but the volume it occupies is certainly not the same. If we characterize this system the same way we characterize the volume, i.e. it by a continuous set of points in a certain *interpretation*, then the discrepancy has a positive definition: at a given moment of time the system coincides with the volume, while at a later moment they do not coincide anymore. However, if we assume that the mass of the system is constant, then the transport can be mathematically described as a spatial process during which this condition is maintained.

Indeed, the condition just stated calls for the vanishing of the time rate of variation of the mass of system. Consequently it is necessary to calculate the time rate of variation of the mass from equation (10.15). In order to represent the time sequence, we have to use the Lie derivative of the differential form under the integral sign along a 'velocity' vector. Thus if we write:

$$\frac{dm_{System}}{dt} = \frac{d}{dt} \iiint_{V(t)} \rho \, dV$$

where V(t) means the fact that the volume is a function of the time sequence, and for the expression of the time rate use the transport equation (6.12), we shall have [(Flanders, 1973); (Betounes, 1983)]:

$$\frac{dm_{System}}{dt} = \iiint_{V(t)} L_{\mathbf{v}}(\rho dx^1 \wedge dx^2 \wedge dx^3). \qquad (10.16)$$

Therefore the velocity field $\mathbf{V}$ is referring to second term of (6.11), and the equation (10.16) will take the form

$$\frac{dm_{System}}{dt} = \oiint_{S(t)} (\rho \mathbf{V}_s) \cdot \hat{\mathbf{n}} \, dA(t) \qquad (10.17)$$

where the surface of integration S(t) is the closed surface delimiting the physical system at the moment 't'. Here the unit normal $\hat{\mathbf{n}}$ orients the element of area of surface S(t) according to the definition $dA(t) \equiv \hat{\mathbf{n}} dA(t)$, and we added the index 's' to the velocity field in order to mark its essential property of being a velocity field only defined on the closed surface S(t) which delimits the system in its evolution.

Physically however, the equation (10.17) is not sufficient: it does not represent the transport in volume but only as a flux through the surface of the physical system, according to its mathematical representation as a set of points. This flux describes only a stationary physical structure, while we are also interested in the physical structural changes of the system during transport. These changes take place only within volume, and physics treats them, in the most general case, as *variations of the density of the system*. Thus, for completeness, from a physical point of view the equation (10.17) must be written in the form

$$\frac{dm_{System}}{dt} = \iiint_{V(t)} \frac{\partial \rho}{\partial t} dV(t) + \oiint_{S(t)} (\rho \mathbf{V}_s) \cdot \hat{\mathbf{n}} \, dA(t).$$

Using here the Gauss theorem in order to express the surface integral by a volume integral, we finally have:

$$\frac{dm_{System}}{dt} = \iiint_{V(t)} \left(\frac{\partial \rho}{\partial t} + \nabla \cdot (\rho \mathbf{v}_s) \right) dV(t) \qquad (10.18)$$

where by $\mathbf{v}$ we mean *the velocity field characterizing the flux of the system as a current*. Let us emphasize once more: the equation (10.18) assumes the space filled with matter, so that in a control volume we have to contemplate only matter, not a physical structure: there is no space there, no 'pores' or 'voids' in the phrasing of Newton. Now, with some usual assumptions of regularity, the arbitrariness of the control volume allows us to declare that this integral equation has a local equivalent:

$$\frac{\partial \rho}{\partial t} + \nabla \cdot (\rho \mathbf{v}) = 0. \qquad (10.19)$$

This is the usual *continuity equation for the mass*.

The most important outcome of this continuity equation is the *Reynolds transport theorem*, which can be obtained if the control volume is decided by a certain physical quantity, say Q (from Quantity!). This theorem arises from the concept of continuity as follows (Reddy, 2008): consider the magnitude 'q' defined by $\rho q \equiv dQ/dV$, which is a *density* indeed, but this time *referred to mass, not to volume*. Instead of equation (10.18), we shall have then

$$\frac{dQ}{dt} = \iiint_{CV} \left(\frac{\partial (\rho q)}{\partial t} + \nabla \cdot (\rho q \mathbf{v}) \right) dV(t);$$

$$Q = \iiint_{CV} \left(\frac{dQ}{dV} \right) dV(t). \qquad (10.20)$$

Expanding appropriately the differential operations under integral sign, we have

$$\frac{dQ}{dt} = \iiint_{CV} \left\{ \rho \left(\frac{\partial q}{\partial t} + \mathbf{v} \cdot \nabla q \right) + q \left(\frac{\partial \rho}{\partial t} + \nabla \cdot (\rho \mathbf{v}) \right) \right\} dV(t).$$

Now, if the continuity equation of the mass (10.19) is satisfied over the region of space contained in the control volume under consideration then we have

$$\frac{dQ}{dt} = \iiint_{CV} \rho \left(\frac{\partial q}{\partial t} + \mathbf{v} \cdot \nabla q \right) dV(t). \qquad (10.21)$$

172

This is the Reynolds transport theorem. In words: if the continuity equation of the mass is satisfied, and the system carries a certain quantity Q attached to the mass in a spatially continuous manner, then *the time rate of variation of this quantity during transport is the volume average of the substantial time derivative of the specific magnitude referred to mass, q ≡ dQ/dm.* The substantial derivative is defined as usual, by equation

$$\frac{Dq}{Dt} \equiv \frac{\partial q}{\partial t} + \mathbf{v} \cdot \nabla q. \tag{10.22}$$

The advantage of working with such specific quantities defined with respect to mass is obvious in the classical cases where we deal with the ideas of force, momentum or kinetic energy and apply them to a continuous system. Before bringing these examples to the fore, let us remind however that this kind of magnitudes are usually *intensive quantities*, in the sense that they do not depend upon the space extension of the physical system they describe, provided this one is a continuum. It is this property that conveys universality to the mathematical relations, thereby giving them even the necessary characteristic of a physical law.

Some Classic Physical Examples

First, consider the second law of the classical dynamics. Written in the usual form of an equation representing the proportionality between force and acceleration, it does not satisfy the requirements connected with the idea of a continuous physical system, whereby the mass itself can vary with time. For a continuous physical system we ought to reformulate the second principle of dynamics, in a new mathematical form accounting for the variation of the mass:

$$\mathbf{F} = \frac{d}{dt}\mathbf{P} \quad \text{with} \quad \mathbf{P} \equiv m\mathbf{v}. \tag{10.23}$$

In words: the time rate of variation of the momentum of a system is provided by the resultant of the forces acting upon that system. Thus, because $d\mathbf{P}/dm \equiv \mathbf{v}$, if the mass is conserved in the volume, the Reynolds transport theorem (10.21) can be written as

$$\frac{d\mathbf{P}}{dt} = \iiint_{CV} \rho \left(\frac{\partial \mathbf{v}}{\partial t} + \mathbf{v} \cdot \nabla \mathbf{v} \right) d^3\mathbf{x}. \tag{10.24}$$

Here $d^3\mathbf{x}$ represents the idea that the volume element $dV(t) \equiv d^3\mathbf{x}$ is taken at the location $\mathbf{x}$ in the volume occupied by the system. The control volume cut from the volume of the system includes also a *control surface* which in this case is a purely material surface. The equation (10.24), which can also be read, for instance, as defining the forces acting upon system by their local action, becomes a little more explicit if we go back to the surface term via Gauss theorem. Using this theorem, we can write:

$$\mathbf{F} = \iiint_{CV} \rho \left(\frac{\partial \mathbf{v}}{\partial t} \right) d^3\mathbf{x} + \oiint_{CS} (\rho \mathbf{v}\mathbf{v}) d^2\mathbf{x}. \qquad (10.25)$$

Here $d^2\mathbf{x}$ is the element of the control surface at location $\mathbf{x}$. This formulation of the second law is important in engineering calculations. However, we consider here only the suggestion it contains: the general way of defining the forces consists of their local action to be recognized in the system both by specific transformations and the specification of their geometrical form, mimicking the geometrical form of that action. Indeed, in equation (10.24) the forces are defined only by their local action upon the system. However, in equation (10.25), we notice that this action takes place in the control volume as well as on the control surface delimiting this control volume. A known case is the Fulton-Rainich theorem (Fulton & Rainich, 1932), which expresses the forces satisfying both *Helmholtz conditions*: $\nabla \cdot \mathbf{F} = 0$ and $\nabla \times \mathbf{F} = 0$, in a continuum. The equation (10.25) is the general form of expressing this very fact: the force $\mathbf{F}$ is geometrically − in fact, topologically − defined, both in volume and in surface, by:

$$\mathbf{F} = \iiint_{CV} (\rho \mathbf{f}) d^3\mathbf{x} + \oiint_{CS} \mathbf{t} \, d^2\mathbf{x} = \iiint_{CV} (\rho \mathbf{f} + \nabla \cdot \boldsymbol{\sigma}) d^3\mathbf{x}. \quad (10.26)$$

Here $\mathbf{f} \equiv d\mathbf{F}/dm$, $\mathbf{t}$ is the traction vector in the control surface, while $\boldsymbol{\sigma}$ is the Cauchy stress tensor defined by those tractions: $\mathbf{t} \equiv \hat{\mathbf{n}} \cdot \boldsymbol{\sigma}$, with $\hat{\mathbf{n}}$ the unit normal of the surface. This unit normal physically orients the surface element $d^2\mathbf{x}$, in the sense of the following relation sometimes even taken as a definition of the Cauchy tensor:

$$\left(\oiint_{CS} \mathbf{t} \, d^2\mathbf{x} \right)^k \equiv \oiint_{CS} t^k d^2\mathbf{x} = \oiint_{CS} \sigma^{kj} n_j \, dA.$$

Using the definition (10.26) in (10.24) while arranging appropriately the terms, results in:

$$\mathbf{0} = \iiint_{CV} \rho \left(\mathbf{f} + \frac{1}{\rho} \nabla \cdot \boldsymbol{\sigma} - \frac{\partial \mathbf{v}}{\partial t} - \mathbf{v} \cdot \nabla \mathbf{v} \right) d^3\mathbf{x}.$$

Again, in reasonable regularity conditions, usually assumed by default, this integral equation can be taken as equivalent to a local equilibrium equation:

$$\frac{\partial \mathbf{v}}{\partial t} + \mathbf{v} \cdot \nabla \mathbf{v} = \mathbf{f} + \frac{1}{\rho} \nabla \cdot \boldsymbol{\sigma} \quad \therefore$$

$$\left(\frac{\partial}{\partial t} + v^k \frac{\partial}{\partial x^k} \right) v^i = f^i + \frac{1}{\rho} \frac{\partial}{\partial x^k} \sigma^{ki}. \qquad (10.27)$$

We wrote the equation in the indicial form also, in order to make the meaning of the differential vector operation clearer. It expresses the fact that the *substantial time derivative of the velocity field* in the structure of a physical system — the material acceleration — is given by both the external forces and the tensions. As we have seen in Introduction, this equation is indispensable to Notttale's theory of the complex fractal fluid.

It is now the appropriate moment for a short digression, which turns out to be useful in defining the place of the wave function in the general economy of concepts of the classical natural philosophy and, in fact, of the natural philosophy in general. As we just said, the equation (10.27) can be regarded as an extension of the second law of dynamics to the material continuum. This fact recommends it as a fundamental equation in the description of a fictitious structure of the matter, whereby it ought to play the very same role as that played by the second law of dynamics for the classical material point. The things are, however, not quite as simple as the classical D'Alembert principle portrays them, because we have not defined yet, up to this point, the way of action of the force **f** in matter. Surely, this way is not the same as the one in which the force acts outside the matter, given, for instance, by the Whittaker-Humbert theory in equation (7.39), which describes the action by a potential involving only the surface of separation between matter and space [(Whittaker, 1903); (Humbert, 1927)]. And even if, in the spirit of Hertz's definition for the material particle, we assume that there are ensembles of such particles, enabling us to give the continuum a structure analogous to a physical structure, the problem of interpretation of stresses and strains still remains. For it is hard, if not impossible sometimes, to ascribe to them inertial properties specific to material particles: the stresses exist *even statically* or, better said, *especially statically*, in which case the inertial properties are practically inexistent, at least at a certain space scale, *where the velocities are comparable with deformation rates*. This is another strong reason why we are compelled to define the Hertz material point statically, and use for it the Wigner's principle expressed by equation (2.19). As we shall see, this is the place where the wave function appears as ontologically necessary: it is an invention of the human spirit at least of the same rank as the forces of Newton.

The results reviewed up to this point are sufficient in order to enable us a discussion on some issues of principle of the natural philosophy, thus bringing to a convenient point some fundamental ideas. That 'convenient point' is the close account of the *transcendence between finite and infrafinite* but from the unique physical point of view of the transport theory.

The Hamiltonian Transport in Finite Volume

Our discussion of this issue starts from the observation that all we have to say is referring to a physical system. This means that the physical system fills the control volume or, more generally, that the control volume defines a part of a physical system. The Reynolds transport theorem given in equation (10.21) is effective if and only if the continuity equation (10.19) is effective, expressing that the mass of that part of the system, which is defined by equation (10.15), is a constant during the evolution along the velocity field **v**. Mathematically speaking, *that part of the system should be completely arbitrary*, in order to allow us to pass from equation (10.18) to (10.19). In fact it is not identified by anything else other than a formal assertion, so that there is no sufficient reason to use the continuity equation, *which is a local statement*. Thus, as a matter of principle, in order that such a statement should not appear quite as rhetorical, the mass *of any possible part* of the system needs to be constant by transport, even the mass of an infinitesimal part, viz. the mass element. This entails the fact that the differential condition (10.5) should also be necessarily satisfied at any location from the volume of the system. In a word, the continuity equation of the mass may not be satisfied, but the transport theorem in the form (10.21) should be certainly true. This can be best illustrated starting from the equation of definition of the momentum, for instance, viz. the equation (10.23). We have

$$\mathbf{F} = \frac{d}{dt} m\mathbf{v} = \frac{dm}{dt}\mathbf{v} + m\frac{d}{dt}\mathbf{v}.$$

If the first term in the right hand side here is vanishing, we are left with only the second term, which leads directly to the expression (10.24) even without the special condition of continuity (10.19). The observation is valid for any physical quantity defined by a density and transported by the physical system, and has fundamental consequences. For, if the elementary mass is conserved in a space process of transport, then this transport must necessarily be Hamiltonian, so that the quantity Q from equation (10.21), for instance, cannot be carried during transport but on an axial vector by vortices, as in equation (10.9), on a vortex by polar vectors, as in equation (10.12), or in every point of the system as in equation (10.14). Let us analyze each of these cases in turn.

In the case of transport along a direction, the transport theorem is of the form (10.21)

$$\frac{dQ}{dt} = \iiint_{CV} \rho \left(\frac{\partial q}{\partial t} + \mathbf{w} \cdot \nabla q \right) dV(t) \tag{10.28}$$

but with the vector **w** along which the transport is done defined by equation (10.9), which we rewrite as:

$$w^k = \varepsilon^{klm} \frac{\omega_{lm}}{\rho}; \quad \omega_{lm} = \frac{\partial v_l}{\partial x^m} - \frac{\partial v_m}{\partial x^l}. \tag{10.29}$$

Here ε is the Levi-Civita symbol. This writing makes the fact obvious that the vector $\mathbf{w}$ is the so called *potential vorticity* from the usual theory of fluids [(Salmon, 1998), Section 4, §§1-4, pp. 197 - 207]. Consequently, if the element of mass is preserved during transport, and if the transport is done along trajectories, the potential vorticity is the only form of physical transport in space. We are talking here of a classical transport obviously, i.e. a transport like that in which the particles hold on to their trajectories, and thus also preserve their identity, which is a normal case if we deal with particles carried on a surface. In a more suggestive 'meteorological' jargon, for instance, along a line the physical quantities are carried by particles in vortices 'swirling like hurricanes'.

In order to make this last observation a little more palpable, assume now the case where the transport is done with the help of a surface: instead of equation (10.28) we must write

$$\frac{dQ}{dt} = \iiint_{CV} \rho \left(\frac{\partial q}{\partial t} + \frac{\nabla \Phi \cdot \nabla q}{\rho} \right) dV(t) \qquad (10.30)$$

because according to (10.12), we have

$$w^{lm} = \frac{1}{2\rho} \varepsilon^{lmk} \frac{\partial \Phi}{\partial x^k} \quad \therefore \quad \mathbf{w} = \frac{\nabla \Phi}{\rho}. \qquad (10.31)$$

where $\nabla \Phi$ is the normal to the carrying surface $\Phi = $ constant. Therefore, in this case we have the conclusion: if the element of mass is preserved during transport, and the transport is done by carrying on surfaces, a vector oriented along the normal to these surfaces, having the magnitude inversely proportional with the density is the only physical form of transport. In other words, in transport on a surface the physical quantities are carried by currents of material particles normal to surfaces.

This was certainly the case when the wave mechanics was born. Indeed, if the quantity Q from equation (10.30) is preserved in the control volume during transport along the vector $\mathbf{w}$, then in reasonable continuity conditions one can assume the vanishing of a rate expressed by a substantial derivative constructed with the help of the vector $\mathbf{w}$, given as in equation (10.31). Thus

$$\frac{\partial q}{\partial t} + \mathbf{w} \cdot \nabla q = 0 \quad \therefore \quad \frac{\partial q}{\partial t} + \frac{\nabla \Phi}{\rho} \cdot \nabla q = 0. \qquad (10.32)$$

This last condition should be compared with that already noticed as necessary in the case of the threedimensional wave theory of Louis de Broglie, namely the definition of the phase and group velocities. Indeed, the equation (10.32) becomes either a condition for the group velocity $\mathbf{w} \equiv \mathbf{U}$ and $q \equiv f$, or a condition for the wave velocity for $\mathbf{w} \equiv \mathbf{u}$ and $q \equiv \Phi$ in the theory of de

Broglie's physical ray. Thus, one can say that, what the classical mechanics has ignored, due to a certain idea of materiality, methodically we should say, namely *the transport on surfaces*, has necessarily forced the occurrence of the wave mechanics. For, the transport of an amplitude is just as physical as the transport of a phase, and it is accomplished through matter too, even if not along trajectories. Let us substantiate these two modes of transport, related in a way or another to the physical concept of surface, by a statistical model of a 'label' and a few epoch-making examples. As to the third mode of transport, we postpone its discussion, insofar as it is directly related to the physical idea of an Einsteinian elevator considered as a Wien-Lummer enclosure, and asks for an examination in a special context.

Transcendence between Volume Element and a Control Volume

It is, incidentally, the case of a note clarifying in some detail the difference between the way in which the labels are conceived in the classical physics and the way we conceive them here. One of the attitudes of the human spirit was to describe an ensemble of material particles — in their instance as classical material points, i.e. mobile positions endowed with physical properties — by properties invariant with respect to 'labelling'. Such an approach in physics is notorious, being forced upon our intellect by the indecision on the dynamics of physical structures. However — fact entirely natural, nonetheless remarkable from our perspective — the physics is enticed to this approach not in connection with the motion of the revolving Hertzian material point in a Kepler problem, but related to the idea of Hertzian material point when *considered as a central body in that problem*, i.e. a body creating the gravitational field. The concept of 'label', as a physical identity carried by a particle, acquires here a bonus of precision: it is now exclusively referring to what we call *Hertz material particle as belonging to a material point*, while in the usual classical acceptance it is referring directly to a Hertz material point.

We shall take the case of Earth as a fundamental example of spatially extended particle. Living in its universe, the man created a science to deal with this environment, from which we can draw some natural conclusions valid at any space scale. That science is geophysics, and we take it now in its instances as meteorology and oceanology, the two sections of geophysics dealing with the fluid parts of the Earth's universe. In these fluid parts of universe, the transport phenomena are the main motivation for study, inasmuch as they do affect our social life directly. The transport of mass, charge and heat in hydrosphere and atmosphere is crucial indeed for our daily social life, and from the study of these transport phenomena we can extract the role of the concept of 'label', and extend it no matter of scale.

A theorem discovered by Hans Ertel in 1942, proved itself subsequently to be the mathematical expression of that invariance to labelling in the case of the physical transport of quantities by molecular ensembles admitting a purely kinematical description [see (Salmon, 1998), pp. 299 - 304]. From our perspective this theorem has a particular importance. In order to reveal that importance, it suffices to give to Ertel's theorem a suggestive statement to be made easily understandable. And that understanding rests upon the simple fact that the Ertel's theorem correlates the two essential transport modes: *the transport preserving the mass element and the transport preserving the mass contained in a control volume.* In order to comprehend this general statement, we shall use here only the argument of Clifford Ambrose Truesdell. And not quite only for the fact that it is the simplest and the most direct argument, indeed, but rather because, alongside, it has the advantage of showing precisely what are the points in which the physics intervenes in order to limit the universality of kinematics (Truesdell, 1951).

That argument starts from the expression of the acceleration field. In the case of continuous systems dealt with in meteorology, this field involves also such collective physical fields as the Coriolis acceleration and the gradient of the square of field velocity, which in the case of fluids describe the turbulence:

$$\mathbf{a} = \frac{\partial \mathbf{v}}{\partial t} + \mathbf{\Omega} \times \mathbf{v} + \frac{1}{2}\nabla(\mathbf{v} \cdot \mathbf{v}). \tag{10.33}$$

The vector $\mathbf{\Omega}$ has the components $\Omega^k \equiv \varepsilon^{klm}\omega_{lm}$ with ω_{lm} defined by equation (10.9). Thus, by this, one implicitly admits a first physical fact, namely that the mass element is conserved when the mass is transported by vortices. Applying the vortex operator in both sides of the equation (10.33) results in

$$\nabla \times \mathbf{a} = \frac{\partial}{\partial t}\mathbf{\Omega} + \nabla \times (\mathbf{\Omega} \times \mathbf{v}). \tag{10.34}$$

Dividing here by the scalar ρ resulting from the definition of the vector $\mathbf{\Omega}$ according to equation (10.9), and expanding alongside the double vector product from the right hand side, finally gives:

$$\frac{\nabla \times \mathbf{a}}{\rho} = \frac{1}{\rho}\frac{D}{Dt}\mathbf{\Omega} + \mathbf{\Omega}\left(\frac{\nabla \cdot \mathbf{v}}{\rho}\right) - \left(\frac{\mathbf{\Omega} \cdot \nabla}{\rho}\right)\mathbf{v}. \tag{10.35}$$

Here for the material derivative we used the definition (10.22). Now, this is the place where the second physical fact intervenes, the first being, as we saw, the condition that the mass element is preserved when the mass is carried by vortices during transport. Namely, in order to write the second term in the equation (10.35) we use the equation (10.19). But that equation is the consequence of a statement referring to the *mass continuity in finite volume*, and thus we must have:

$$\frac{\nabla \cdot \mathbf{v}}{\rho} = \frac{D}{Dt}\left(\frac{1}{\rho}\right) \qquad (10.36)$$

expressing the fact that the divergence of the velocity field of the fluid continuum is given by material derivative of the logarithm of fluid density. Thus, the equation (10.35) becomes:

$$\frac{D}{Dt}(\mathbf{w}) = \frac{\nabla \times \mathbf{a}}{\rho} + (\mathbf{w} \cdot \nabla)\mathbf{v}. \qquad (10.37)$$

This equation can be further generalized (Truesdell, 1951a), and its generalisation is actually the key of Hans Ertel's observations from 1942. Namely, if ψ is a quantity conserved by transport in finite volume, in the sense that its material time derivative is null, then the material time derivative of $(\mathbf{w} \cdot \nabla\psi)$, calculated in the particular time sequence in which the conservation law of ψ takes place, has the meaning of a density:

$$\frac{D}{Dt}(\mathbf{w} \cdot \nabla\psi) = \frac{1}{\rho}\frac{\partial(\rho, 1/\rho, \psi)}{\partial(x^1, x^2, x^3)}. \qquad (10.38)$$

Here 'p' is the pressure, a physical variable which seems quite natural in all of the classical cases. In order to prove this equality, Truesdell takes notice of the fact that there is a general identity

$$\frac{D}{Dt}(\mathbf{w} \cdot \nabla\psi) = \frac{\nabla \times \mathbf{a}}{\rho} \cdot \nabla\psi + (\mathbf{w} \cdot \nabla)\left(\frac{D\psi}{Dt}\right) \qquad (10.39)$$

deriving from (10.37) by scalar multiplication with $\nabla\psi$, where ψ can be a scalar, a component of a vector, or even an entry of a tensor of arbitrary dimension. Now, if ψ is materially conserved, then (10.39) reduces to

$$\frac{D}{Dt}(\mathbf{w} \cdot \nabla\psi) = \frac{(\nabla \times \mathbf{a}) \cdot \nabla\psi}{\rho}. \qquad (10.40)$$

The mathematical result known under the name of Ertel's theorem, can then be expressd as:

$$\frac{D}{Dt}(\mathbf{w} \cdot \nabla\psi) = 0. \qquad (10.41)$$

The usual way of proof − which is actually the way Hans Ertel himself followed (Ertel, 1942) − is by passing from equation (10.39) to (10.38) via *constitutive considerations* relating to the fluid in motion (Truesdell, 1951a), and then admitting a certain behaviour of the fluid, whereby the Jacobian from the right hand side of the equation (10.38) vanishes. The reason for this is based on the fact that the constitutive relationship which involves the pressure and density in the relation expressing the vortex of acceleration:

$$\nabla \times \mathbf{a} = \nabla p \times \nabla(1/\rho) \tag{10.42}$$

shows that the equation (10.40) goes into (10.38). Thus, if a functional relation exists between pressure and the density the equation (10.41) follows.

Truesdell insists, however, on a fundamental conclusion regarding Ertel's theorem: it can be obtained only by geometrical considerations following exactly the line above, *without involving any constitutive considerations*. In this, one uses vector calculus, assisted only by elementary mass conservation and continuity equation, with no concept of pressure, which necessitates further elaboration in order to be physically explained. To wit, because $\nabla\psi$ is a gradient, we have the vector identity $(\nabla \times \mathbf{a}) \cdot \nabla\psi \equiv \nabla \cdot (\mathbf{a} \times \nabla\psi)$ which, when used in (10.40) produces a purely kinematical result, obviously universal: the Ertel's theorem is valid without any supplementary physical condition − constitutive relation, the definition of pressure etc. − *when the acceleration field is oriented along the normal at the surface* ψ = constant. In fact, one can extract a much more general conclusion, in the sense that it does not depend on the fact that ψ is materially constant or not (Truesdell, 1951b), just starting from the equation (10.39). However, this conclusion takes shape either in the nonhomogeneous differential form:

$$\frac{D}{Dt}(\mathbf{w} \cdot \nabla\psi) = (\mathbf{w} \cdot \nabla)\left(\frac{D\psi}{Dt}\right)$$

or only 'as a mean', as Truesdell expresses it in the work just cited:

$$\iiint_{Volume} \rho\left[\frac{D}{Dt}(\mathbf{w} \cdot \nabla\psi) - (\mathbf{w} \cdot \nabla)\left(\frac{D\psi}{Dt}\right)\right] dV = 0$$

and only under the condition that the acceleration field is oriented along the normal to surface ψ = constant, in any of its points.

It is therefore highly significant, from the point of view of the theory of ensembles of Hertz material particles, which possibly casts the continuum into a primary structure − the interpretative structure − that the magnetic motion of the Poincaré type, generated by the force (6.50), occurs only on conical surfaces. First, on this type of surfaces the geodesics are 'magnetic trajectories' so to speak, i.e. trajectories along which the acceleration is oriented along the normal to surface. By its definition, the geodesic on a surface is a curve whose normal coincides with the normal of surface [(Struik, 1988), p. 131]. Thus, the ensemble of Hertz material particles from the nucleus of the material point creating the gravitational field in the Kepler problem, for instance, can surely be described via Ertel's theorem *at any space scale*, for that theorem is independent of any constitutive relationship. We can thus realize that this theorem according to Truesdell, *independent of any classical physical structure* − is appropriate to serve in describing even that fictitious structure 'confined' in the nucleus of every material point − in the very same manner the quarks and partons are thought to be confined in matter − insofar as it does not assume constitutive relations which would imply the idea of a physical structure. The Ertel's theorem assumes only a continuous behavior of the mass at any space scale of the matter continuum and, as an aside hypothesis, only a specific orientation of the acceleration field. Or, according to the regularization procedure in its focal alternative (Burdet, 1968, 1969), these conditions are always satisfied upon conical surfaces having the base a spherical cap, i.e surfaces defining the usual solid angle. It is this property which, in our opinion, confers to Ertel's theorem a great gnoseologic importance.

As an incidental general observation, in spite of its great methodological importance, it seems that the Ertel's theorem found hardly its way within the scientific observance of the last century. It is only after a decade from its discovery [see (Ertel, 1942); for an English translation of the essential works of Hans Ertel, one can refer to (Schubert, Ruprecht, Hertenstein, Nieto Ferreira, Taft, Rozoff, Ciesielski, & Kuo, 2004)] the true value of this theorem started to be noticed as such, through the works of Clifford Ambrose Truesdell (Truesdell, 1951). For the needs of present work, we should rather recommend the presentation of Rick Salmon, which seems clearer, as being intuitively more explicit than many others (Salmon, 1988, 1998). Besides, it deliberately insists upon the fact that the theorem is the implicit expression of a *classical invariance to labelling*, here present in the equation (10.33) via the Jacobian expression. The idea is as attractive as it gets, in view of the fact that, as we shall see shortly, the non-Euclidean geometry of essential physical quantities in a Hertz material point − the gravitational mass and the two charges, magnetic and electric − defines the density of the matter ensemble of particles by a Jacobian.

As a general conclusion of the preceding analysis, a physical image of the continuum depends upon two necessities rationally enacted by mathematics during the process of interpretation. These occur as a natural consequence, in view of the idea of ensemble of material particles − the essential tool of interpretation. First, one has the conservation of the mass element, defining the density in Newton's manner, *with reference to a volume element*. Then, one has the idea of conservation in any control volume, expressed by the transport theorem, requiring in turn *the physical content of a finite volume*. We thus have shown that the physical realization of the first necessity is sufficient in order to warrant the possibility of the physical realization of the second, via the Reynolds transport theorem. The physical conservation of mass element is realized in three ways: two of them are referring to the transport on and between surfaces, and are represented by the equations (10.9) and (10.12) which define it mathematically. As far as the third way is concerned, the one characterized by equation (10.14), this raises the problem of interpretation of the quantities V^{ijk} and it has a particular importance in the history of physics. We postpone these issues for a later specific elaboration in this work.

Chapter 11. A Hertz-type Labelling in a Madelung Fluid

The first in our order of reason here, arises the problem of expressing coordinates in the general circumstances representing the *arithmetization of a continuum*. One of the natural ideas in the spirit of Nicholas Georgescu-Roegen, would be to replace them by specific densities of statistical distributions (Calmet & Calmet, 2004). It is known that the families of these densities can be organized as a Riemannian space [see, for instance, (Burbea, 1986)], having a metric tensor given by the *Fisher information measure*:

$$g_{\mu\nu}(\mathbf{s}) = \int_{\mathbf{x}} d^4\mathbf{x} \frac{1}{p_{\mathbf{x}}(\mathbf{x};\mathbf{s})} \frac{\partial p_{\mathbf{x}}(\mathbf{x};\mathbf{s})}{\partial s^{\mu}} \frac{\partial p_{\mathbf{x}}(\mathbf{x};\mathbf{s})}{\partial s^{\nu}}. \qquad (11.1)$$

Jacques and Xavier Calmet operate in the four-dimensional space of the general relativity, so that the formula given here has the basic features of original work, specifically $\mu, \nu = 0,1,2,3$ or $1,2,3,4$ depending on the convention of use of coordinates and time. However, from a statistical point of view it should be valid in any dimension. Explanation of the symbols is as follows: the vector $\mathbf{x}$ is the position in the space $\mathbf{X}$ of the stochastic event position-time, while the vector $\mathbf{s}$ represents the four parameters of the probability density $p_{\mathbf{x}}(\mathbf{x};\mathbf{s})$. Although in a very specific expression, we have here one of the properties that allowed the first conclusions regarding the necessity of probabilistic interpretation of the wave function: the coordinates in which the metric (11.1) is understood *are parameters of distributions which characterize them theoretically*. In other words, the Riemannian geometry generated by (11.1) is actually a geometry in the *parameter space of the statistics of coordinates*.

That property descends from a classical case which, in our opinion, becomes critical with Laurent Nottale's views. As a rule, however, this case is contemplated in some other circumstances regarding the fundamental physics. Here, however, it can be illuminating in establishing the general view on the parameters of a statistical distribution. The classical case in question is concretely represented by the *Ehrenfest theorem*, according to which the classical quantities are ensemble averages of the microscopic quantities of corresponding connotation. The property proves indeed to be general, but with the generality well delineated from a statistical point of view: *every classical quantity is represented by certain parameters of the families of statistical distributions describing the geometry of a space-time.* For, indeed, in the present context, the traditional task of Ehrenfest's theorem lays on it the aspect of just a particular case among many possible others: it is that case where the *means and standard deviations* (or the variance matrices in general) *serve as parameters for the distributions over ensembles of microscopic quantities.*

This would mean, for instance, that even a classical trajectory must be statistically characterized, and this description is quite precise: practically speaking, not every position is *measurable* along trajectory, as the classical mechanics seems to claim. The phase space of a classical material point should be indeed a continuum by the mathematical rigor. The act of measurement, however, *chooses from this continuum only means* or, in the most general case, *parameters related to these means* by some mathematical relationships based on probability densities. The problem here is to recognize that, generally, the coordinates describing the phase space thus characterized are not exclusively means or exclusively standard deviations, but still some other quantities of physical interest, liable though to connections with those traditional particular parametrizations. This is the essence of equation (11.1).

Fact is that if in equation (11.1) we write the probability density in a three-dimensional space in a form suggesting a Born-like interpretation of the wave mechanics:

$$p_{\mathbf{x}}(\mathbf{x}; \mathbf{s}) = [\psi_{\mathbf{x}}(\mathbf{x}; \mathbf{s})]^2 \qquad (11.2)$$

a few things start to unravel. In the spirit of Louis de Broglie's idea, this would mean that the wave function itself is considered an 'optical' amplitude, physically interpreted by an ensemble of Hertz material particles. Then the Fisher information measure is given by the metric tensor

$$\mathbf{g}(\mathbf{s}) \equiv 4 \int_{\mathbf{x}} d^3\mathbf{x} \cdot [\nabla_{\mathbf{s}} \psi_{\mathbf{x}}(\mathbf{x}; \mathbf{s}) \otimes \nabla_{\mathbf{s}} \psi_{\mathbf{x}}(\mathbf{x}; \mathbf{s})]. \qquad (11.3)$$

Here $\otimes$ signifies the dyadic product of the vectors. Indeed, the writing in equation (11.2) just suggests the idea of wave function: if the three-dimensional parameter $\mathbf{s}$ is a position vector in the usual space, then the equation (11.3) represents the tensor generalization of the functional used by Erwin Schrödinger in his variational principle generating the stationary equation for the free particle at rest (Schrödinger, 1933). That functional is obviously more restrictive, being in the present context just the trace $tr(\mathbf{g})$ of the metric tensor, up to a constant numerical factor. Submitted to the variational principle used by Schrödinger, it gives the usual Laplace equation:

$$\nabla_{\mathbf{s}}^2 \psi_{\mathbf{x}}(\mathbf{x}; \mathbf{s}) = 0 \qquad (11.4)$$

and its interpretation here raises a few problems, opening nevertheless the way to a special interpretation of the facts.

There is a solution of equation (11.4), satisfying almost all of the requirements of the discussion right above. It can be written in the form

$$\psi_{\mathbf{x}}(\mathbf{x};\mathbf{s}) = (1 - 2\mathbf{s}\cdot\mathbf{x} + \mathbf{s}^2\mathbf{x}^2)^{-1/2}. \qquad (11.5)$$

The chief one among these requirements, is obviously the fact that such a wave function has an algebraical structure in which the parameters $\mathbf{s}$ and the position variables $\mathbf{x}$ *enter on equal footing* so to speak. This would represent the natural property of a mean, namely that it should be, quantitatively speaking, *of the same order of magnitude as the values from the ensemble* it describes, possibly even one of those values. In the long run, this property means that statistics preserve the scale of things, and the ideas of Nicholas Georgescu-Roegen are indeed a sound basis for the mathematics involved in theoretical statistics. In this regard, the function (11.5) can very well be considered as a solution of the Laplace equation in the position variables $\mathbf{x}$, and in fact this can add one more important property to the characterization of 'equal footing' of $\mathbf{x}$ and $\mathbf{s}$. The function $\psi_{\mathbf{x}}(\mathbf{x};\mathbf{s})$ would then give a probability density which, according to (11.2) is, up to multiplicative constant, also a symmetric function in $\mathbf{x}$ and $\mathbf{s}$:

$$p_{\mathbf{x}}(\mathbf{x};\mathbf{s}) = (1 - 2\mathbf{s}\cdot\mathbf{x} + \mathbf{s}^2\mathbf{x}^2)^{-1}. \qquad (11.6)$$

Then the metric tensor (11.3) can be written as a mean, for the integrand in (11.3) is the function

$$p_{\mathbf{x}}(\mathbf{x};\mathbf{s}) \cdot [\nabla_{\mathbf{s}}\ln p_{\mathbf{x}}(\mathbf{x};\mathbf{s}) \otimes \nabla_{\mathbf{s}}\ln p_{\mathbf{x}}(\mathbf{x};\mathbf{s})]$$

$$\equiv \frac{4}{1 - 2\mathbf{s}\cdot\mathbf{x} + \mathbf{s}^2\mathbf{x}^2} \cdot \frac{\mathbf{x} - \mathbf{s}\cdot\mathbf{x}^2}{1 - 2\mathbf{s}\cdot\mathbf{x} + \mathbf{s}^2\mathbf{x}^2} \otimes \frac{\mathbf{x} - \mathbf{s}\cdot\mathbf{x}^2}{1 - 2\mathbf{s}\cdot\mathbf{x} + \mathbf{s}^2\mathbf{x}^2}. \qquad (11.7)$$

In other words, the metric tensor (11.3) is the average of the dyadic product with itself of the vector

$$\mathbf{m} = \frac{\mathbf{x} - \mathbf{s}\cdot\mathbf{x}^2}{1 - 2\mathbf{s}\cdot\mathbf{x} + \mathbf{s}^2\mathbf{x}^2}. \qquad (11.8)$$

The notation $\mathbf{m}$ is intended to suggest the idea of a *mean*, because we have to do indeed with a mean in the sense to be presently described.

First note that the vector (11.8) can be written as

$$\mathbf{m} = \nabla_{\mathbf{s}}F(\mathbf{x};\mathbf{s}), \quad F(\mathbf{x};\mathbf{s}) \equiv \ln\psi_{\mathbf{x}}(\mathbf{x};\mathbf{s}) \qquad (11.9)$$

which means that $\mathbf{m}$ is an average position of an ensemble of positions having $F(\mathbf{x};\mathbf{s})$ as a *partition function*. According to this interpretation, one can say even what kind of ensemble is this one, from the point of view of the very theory of probabilities: it is an ensemble having as probability density a *natural*

186

exponential with quadratic variance function in three statistical variates locating a position. [see (Letac, 1989); (Casalis, 1996), for the concept and general properties]. Such an ensemble is a genuine concept of quantum physics, forasmuch as this one was constructed from its very beginnings based on it (Mazilu, 2010). Indeed, Planck's quantization is the very first such example in physics, but referring to a single statistical variate, the energy density of the thermal radiation field. Here, however, we have to do with three statistical variates, and this needs a little introduction into subject, with a few definitions.

A set of elementary probabilities, depending on three parameters arranged as a vector **s** and expressed in the form

$$P(\mathbf{x}; \mathbf{s}, \mu)(d\mathbf{x}) = \exp\{\mathbf{s} \cdot \mathbf{x} - k_\mu(\mathbf{s})\} \cdot \mu(d\mathbf{x}) \qquad (11.10)$$

is an exponential family with three parameters. Here $\mu(d\mathbf{x})$ is a certain *a priori* measure of space, and the index μ is given to function 'k' in order to show the dependence of density normalization on the *a priori* measure of space. The mean vector and the variance matrix of such an ensemble are given by the expressions:

$$\nabla_{\mathbf{s}} k_\mu(\mathbf{s}) = \int \mathbf{x} \cdot P(\mathbf{x}; \mathbf{s}, \mu)(d\mathbf{x}) \equiv \mathbf{m};$$

$$(\nabla_{\mathbf{s}} \otimes \nabla_{\mathbf{s}}) k_\mu(\mathbf{s}) = \int (\mathbf{x} - \mathbf{m}) \otimes (\mathbf{x} - \mathbf{m}) P(\mathbf{x}; \mathbf{s}, \mu)(d\mathbf{x}) \equiv \mathbf{V}. \qquad (11.11)$$

Certainly, the first one of these equations can be identified with (11.9) for $k_\mu(\mathbf{s}) \equiv F(\mathbf{x}; \mathbf{s})$, but there is a problem: the mean over our ensemble depends on the elements of the ensemble itself. This is not a regular occurrence, at least from an orthodox statistical point of view. However, from a certain physical point of view, we can relegate it to a known issue of the wave mechanics, and this allows for an alternative interpretation of statistics. Namely, in view of the first relation from (11.11), the equation (11.9) can be written as

$$\psi_{\mathbf{x}}(\mathbf{x}; \mathbf{s}) = \exp\{k_\mu(\mathbf{s})\} \qquad (11.12)$$

so that the wave function is liable to be further defined by some *hidden parameters* besides those represented by the vector **s**. These parameters are contained in the *a priori* measure $\mu(d\mathbf{x})$ of the space.

Before any further discussion on this issue, let us consider the second equation from (11.11). Regardless of the existence of the hidden parameters, the variance matrix **V** can be written exclusively in terms of the vector **m** in the form which, following Gérard Letac, describes a family of natural exponentials with three parameters, having quadratic variance (matrix) function:

$$\mathbf{V}(\mathbf{m}) = \sum_{i,j=1}^{3} \mathbf{A}_{ij} m_i m_j + \sum_{k=1}^{3} \mathbf{B}_k m_k + \mathbf{C}, \qquad (11.13)$$

where $\mathbf{A}_{ij}, \mathbf{B}_i$ and $\mathbf{C}$ are symmetric matrices. Now, it is quite obvious that our equation (11.10) with $k_\mu(\mathbf{s})$ defined by (11.12) fits into this definition, for we have after a short calculation based on the second equation (11.11), the following covariance matrix:

$$V_{kl}(\mathbf{m}) = (\mathbf{m})^2 \delta_{kl} - 2m_k m_l. \qquad (11.14)$$

This quadratic matrix is obviously of the form (11.13) for $\mathbf{B}_i$ and $\mathbf{C}$ all null matrices, and $\mathbf{A}_{ij}$ given by the fourth order orthogonal matrix

$$(\mathbf{A}_{ij})_{kl} = \delta_{ij}\delta_{kl} - \delta_{ik}\delta_{jl} - \delta_{jk}\delta_{jl}. \qquad (11.15)$$

Thus, there is nothing in this scheme of definition of the multiparameter natural exponentials with quadratic variance matrix function, which would not befit our vector $\mathbf{m}$ in its capacity of a mean, as defined by equation (11.8). Except, perhaps, a genuine representation of what has come to be called *statistical sufficiency*. Within the present geometrical theory, this idea can be best understood as follows.

One can notice that the function defined by equation (11.6) has indeed one of the essential properties of a probability density function: it is positive over the whole *a priori* range of either $\mathbf{x}$ or $\mathbf{s}$. This can be easily verified, just by writing the quadratic function from the denominator of (11.6) in spherical polar coordinates (r, θ, φ), and referring the colatitude angle θ either to the direction of the parameter vector s or to the direction of position vector x, as the case may happen to be convenient. Thus we can write:

$$1 - 2\mathbf{s} \cdot \mathbf{x} + \mathbf{s}^2 \mathbf{x}^2 \equiv 1 - 2sr\cos\theta + s^2 r^2.$$

This quadratic polynomial — either in the variable 'r', or in the variable 's' — is always positive, because its discriminant is unconditionally negative. Consequently, the function $p_\mathbf{x}(\mathbf{x}; \mathbf{s})$ is always positive, no matter if considered as a function of $\mathbf{x}$ with the parameters $\mathbf{s}$, or a function of $\mathbf{s}$ with parameters $\mathbf{x}$.

Problems arise only with the normalization, a second essential condition that needs to be satisfied by a probability density. If this condition is not satisfied, the Fisher measure of information cannot be defined as such, and therefore the metric tensor cannot be defined by equation (11.1). And here it is quite obvious that the integral of the function (11.6) over the whole a priori range of the vector $\mathbf{x}$, is not a finite quantity, as it should be in order to characterize a continuous probability. More to the point, that integral can be written as

$$\int_0^{2\pi} d\varphi \int_0^\pi d\theta \cdot \sin\theta \int_0^\infty \frac{r^2 dr}{1 - 2sr\cos\theta + s^2 r^2}$$
$$\equiv \int_{-1}^{+1} d\eta \int_0^\infty \frac{r^2 dr}{1 - 2sr\eta + s^2 r^2} \qquad (11.16)$$

188

where $\eta \equiv \cos\theta$. A second integration over η gives the value:

$$\frac{2\pi}{s} \int_0^\infty r\,dr \cdot \ln\frac{|1+sr|}{|1-sr|}. \qquad (11.17)$$

The final integration can be done by regular means, for there is a primitive of the integrand in this equation [see e.g (Gradshteyn & Ryzhik, 2007), ex. 2.729.2]. However, the result is not always positive and, moreover, even if positive it is not always finite.

One can only produce a meaningful result in real numbers under the constraint $-1 \le \lambda r \le 1$ that guaranties the right values of the modulus in (11.17). However, because 'r', as well as 's' for that matter, must be always positive, it is clear that in order to make sense, the integrand needs an even more restrictive constraint: $0 < \lambda r < 1$. Thus, 'r' can take any positive value, as the integration in (11.16) requires, but not unconditionally: it depends on the parameter 's' that the integral makes sense and if so, this happens only over a limited interval of values of 'r'. Any way we look at it, the integral cannot be calculated in finite terms over the whole range of *a priori* values of the radial coordinate, and consequently the function $p_\mathbf{x}(\mathbf{x};s)$ cannot be normalized as such.

The things seem to get in order *within physics* only if we conceive the Fisher information measure in the spirit in which it was conceived first, i.e. as a test functional of the *physical homogeneity* of the data sampled from a given population (Fisher, 1925). The physical homogeneity of a population with respect to a given physical quantity does not mean only a *significant grouping of the results* of measurements of that quantity, but also *the same probability density used in estimating* by sampling the statistics of that grouping (mean vector and variance matrix). If these two conditions are not satisfied the sampling does not produce efficient, viz. reliable results from a theoretical point of view. The Fisher information measure was actually first conceived as *a measure of efficiency*. As it turns out, the indecision in the probability measure is correlated with the indecision in the evaluation of a physical quantity as a statistic of the population under sampling (Frieden, 2004). Our problem is then that of getting down to considerations of the relationship between the probability density and physical measurement of quantities, and this can only be done if we have a *sampling process* producing the evaluations from equation (11.11), and involved in some physics related to the probability density from equation (11.6).

A first question would be if *the limitation* in the *a priori* range of space variables produces meaningful probability densities. The answer is only conditionally affirmative, in that it is contingent upon the possibility of producing true probabilities. For instance, in the case above, under our scrutiny here, we are referring only to coordinates of positions in some range of space extensions. Let us therefore assume that *we cannot measure but a limited range of the radial coordinate*, say between 0 and R for any given 's'. Then we can define 's' by the condition of a fixed $\xi \equiv \lambda R < 1$. This condition is equivalent with admitting that we cannot talk of a sampling which would justify the definition of Fisher information, but only for a limited space range inside a sphere of radius ξ/R. This certainly looks like a sampling defined primarily by the radius of sphere. In this case we have indeed, instead of the undecided value of equation (11.16), the finite value calculated as in the following equation:

$$\frac{2\pi}{s} \int_0^\infty r dr \cdot \ln\frac{|1+sr|}{|1-sr|} = R^3 \frac{2\pi}{\xi^3} \left(\xi + \frac{1-\xi^2}{2}\ln\frac{1-\xi}{1+\xi} \right) . \qquad (11.18)$$

The right hand side here can very well be taken as a normalization factor of the probability density (11.6), obviously if it is positive. As it happens, this quantity is always positive within the given conditions, as can be checked by direct calculations. The sphere of radius R is therefore a proper tool which can be used to characterize such a space sampling process: *one samples only points inside this sphere*. The problem is now transferred to the physical level: *what is the physical meaning of such a sampling*. As it turns out, this physical interpretation is not quite out of hand from the point of view of the measurement, if we reconsider the equation (11.8).

Fact is that the equation (11.8) embodies a succession of transformations *involving a sphere*. First, we have a transformation by "reciprocal radii" as they say, i.e. an inversion proper with respect to a sphere that can represent the scale of things contained inside it. In order to get an idea of what is going on here, let us say that we need to pinpoint the Sun's position. Now, *we know scientifically* that the Sun is a star, and the idea of locating a star is very simple, since a star can be assimilated with a point, therefore it has a first-hand geometrical position in a certain reference frame. However, this characterization is valid *only at a certain space scale*. We are aware of this fact by the very same science that allows us to place the Sun and stars under the same concept. For if they are physically describable in the same manner, then any star should also have the same properties which the Sun has. The first property that jumped the human mind in this particular problem, was that a star should have a planetary system like the Sun itself. However, this property is not the one to be theoretically considered first. And this not because the evidence for such an occurrence is not a first-hand evidence and, even as such, too scanty. By far more important, from a physical point of view, is that *a star should have spatial extension*, like the Sun has. Now, if a star has a spatial extension, then its precise location is a matter of scale of proportion of the space we can reach by our physical means. Indeed, the Sun cannot be precisely located, inasmuch as it has an obvious finite extension in space. When we talk of the Sun's position we really mean the position of a point within the space occupied by Suns physical structure. Approximating Sun's shape with a sphere, we may accept, for instance, that its position is the position of the center of that sphere. In order to maintain the very same philosophy for all the stars, we then need to characterize their space extension by a parameter, or a combination on a set of parameters, allowing us to say that only in a certain limit of these parameters a star is *sufficiently characterized* by a position. The Sun − and so a star in general − is surely *not sufficiently characterized* by a position in space.

Before entering deeper into the *idea of sufficiency*, let us continue with characterization of the equation (11.8) the way we mentioned, i.e. as a succession of geometrical transformations, because this way the idea of relation between space scale and position might be easier to comprehend. The inversion proper can be represented by the transformation

$$\mathbf{x} \quad \rightarrow \quad \mathbf{R}^2 \frac{\mathbf{x}}{\mathbf{x}^2}. \tag{11.19}$$

Here R is the radius of the sphere accomodating the inversion, and the positions are considered in an Euclidean reference frame with origin in the center of this sphere. The point having the position $\mathbf{x}$, and its transformed, $\mathbf{x}'$ say, are located along the same radius of the sphere, one inside the other outside it. But this topology heavily depends on the specific details of the physical problem whose geometric solution we need. An example related to our discussion from the present work will better explain what should be understood by 'specific details', helping us, at the same time, in reducing these details to essentials.

Assume that the sphere accomodating the inversion is a light wave surface. If the radius we are talking about is the longitudinal axis of a physical light ray, we might ask ourselves what is happening with the other rays defining its physical structure (Mazilu & Porumbreanu, 2018). Thus, the transformation (11.19) should be true indeed, but only for the axis of physical ray: the other rays − the mathematical rays of this physical structure − behave differently, depending on their position within the physical ray at the point where they touch the sphere. This is the well-known property of a spherical diopter, a concept usually taken as the basis of all of the constructions of geometrical optics. In general, we may assume that a physical ray behaves in such a way that its constitutive geometrical rays can be properly considered as radii with respect to the given sphere *only from some displaced points*. This may be taken as a particular way of expressing the idea that reflection and refraction phenomena are dictated by the general position of the physical ray with respect to the normal to the wave surface, a sphere in this particular case. Be it as it may, the displaced positions will then depend on three parameters:

$$\mathbf{R}^2 \frac{\mathbf{x}}{\mathbf{x}^2} - \mathbf{p} \tag{11.20}$$

where $\mathbf{p}$ is a vector that may depend on time or on the details of the physical system to be thus described, which, in the case here considered is our physical ray. Now, applying again an inversion to the position (11.20) in order to bring it within the same side of the sphere as the original position — i.e. inside or outside — gives the position from equation (11.8), with

$$\mathbf{s} \equiv \frac{\mathbf{p}}{R^2}. \tag{11.21}$$

Therefore, $\mathbf{m}$ describes — here, as well as presumably everywhere — an ensemble of positions generated by starting from a position $\mathbf{x}$, through the radii of a physical ray in the manner just explained. The word 'presumably' may not illustrate the right stand to be taken in this kind of description, for, as far as we are aware, it works everywhere, provided we have a proper interpretation of the parameters of the transformation (11.8). To wit, it is understood here as meaning that we just did not cover all of the situations to which these reasons apply, and may exist some situations of which *we* simply are not aware. However, the wave mechanics as conceived by Louis de Broglie, can be cited as a case illustrating the reasoning, *provided the wave function is defined by Schrödinger's time-dependent equation.* Such a conclusion may, at least as far as we are concerned, obliterate by its grandeur any other possibilities of thinking, so that we feel as having to express it with a certain hesitation. Hence that 'presumably'!

Strange as it may seem, this example does not concern the wave mechanics exclusively. It is the very root of the modern physical optics and, taken from the perspective of our presentation here, it shows, again, that the human knowledge is utterly continuous: there are no 'quantum leaps' of knowledge! Indeed, the modern physical optics started by adding the difraction phenomenon to the old phenomenology involving just reflection and refraction of light (Fresnel, 1827). This completion of phenomenology triggered the necessity of a closer considera- tion of the idea of infrafinite in optics, so that the differential geometry started to be involved routinely in the construction of the wave surface [see (Hamilton, 1841) for a clear geometrical illustration of the procedure of construction of the Fresnel wave surface]. Thus, the Fresnel construction of the wave surface is, in the field of physical optics, a first example of passage from infrafinite to finite and vice versa. The wave mechanics only added a physical interpretation to this objective process. However, the wave mechanics also brought in something new, which comes to our mind if we ask what is the general physical procedure of constructing a physical light ray. Then it becomes clear the fact that, in unfolding the idea of Nicholas Georgescu-Roegen, the choice of an origin is just as important as a gauging in the definition of a reference frame. These are involved in the construction of the frame exactly the way in which the origin of sphere and the sphere of inversion are involved in the inversion as described above. Only, as the wave mechanics teaches us, the sphere is to be replaced by a general surface, in the manner of Humbert for instance, while the a priori choice of origin has to be replaced by a choice according to a physical process of propagation, or an equivalent procedure.

An example a little different in its nature, may be helpful in bringing us to a better appreciation not only of the general geometry, in its Riemannian metric form, but also of the physics involved here. Speaking of the same physical ray of light, we need to consider the human eye, and the idea of physiological mechanism of vision: after all, this is physics too! Thus a physical ray is concentrated by the diopter of eye, in the way just described by the inversion above, on the retina, but over a region given by such images and quantitatively characterized by a statistics like 'the mean' from equations (11.11). These regions ought to be described as ensembles (Hoffman, 1966), and their geometry is the metric geometry of a **sl**(2,R) group algebra [see also (Resnikoff, 1974) for a general theory of vision colors].

Therefore the probability density defined by equation (11.2) starting from (11.5) is only correct, even *a priori*, on limited ranges, not over the entire space as it would seem necessary. This condition can be endorsed within *Hertz's natural philosophy* where a clear probabilistic interpretation is then necessary. Namely, a position in space cannot be assigned but with a limited degree of precision, and only through the intermediary of a material point, as the man does not have any other possibility. The Hertz material particles, the only ones 'entitled' to assign a position in space, are out of our reach! In view of this, it becomes necessary to assign a first *a priori* geometrical measure to the space ranges we need in order to locate a certain position. The above statistical approach of the conformal transformation provides a way for such a construction, through the idea of *sphere of inversion.*

Again, an example infused with a little physics should be illuminating here. Assume indeed, that we are to assign positions to the *centers of force* of the stars from a neighborhood of our universe. We choose here 'the centers of force' in order to suggest the idea that a position is always to be assigned to a physical thing, but that very physical thing may not be directly accessible. For instance, in the case of Sun, we may assign a location for its geometrical center, but unfortunately that is only a geometrical thing. The center of force of the Sun's gravitational field can be located everywhere inside the matter of Sun, or in fact even outside it, as our presentation of the Kepler problem shows. Its position depends, even in the ideal case, on the attracted point, to say the least. In this respect the position of the center of force may also depend on the position of the attracted point on its orbit. Fact is that, in general, a physical position may not be accessible as such, and this is the point where the probability enters the physical knowledge.

If the stars from our chosen system, i.e. the stars which are... suns, are all of the same dimensions and shape — let us say spheres of radius R_0 — *we may* be able to locate their geometrical centers in an arbitrary reference frame.

Assuming then that the system is governed by Newtonian gravitation, *we may* further be able to locate to a certain extent even the centers of forces of every one of the stars. That 'we may' here, expresses a lot of doubt, usually left aside in actual researches. In order to show what is left aside, let us see in what conditions 'we may' be able to establish the center of force of a certain star from our system. Let us choose a star from the system and follow the following procedure for discovering the center of force. Using the reference frame with its origin in the geometrical center of the chosen star, locate three other different stars of the system, for instance in three different directions in space. The Newtonian force exerted by our star upon the other stars, will be of the form $\kappa \mathbf{r}/r^3$, where $\mathbf{r}$ is the position of that other star. Now, as long as κ is a constant, the following observation is true: every component of this Newtonian force is a solution of the Laplace equation in the chosen reference frame, in Cartesian coordinates. Then by the Kelvin's inversion theorem, the force can be considered as the inverse of the position of certain point with respect to the sphere representing the chosen star.

Perhaps it is better to insist on this point, for it is of importance. Take one component of Newtonian force, say $\kappa x/r^3, r^2 \equiv \sum(x^2)$; if κ is a constant, then this component is a solution of Laplace equation in the Euclidean coordinates (x, y, z), i.e. it is a harmonic function. Now, the Kelvin inversion theorem tells us that if $V(\mathbf{r})$ is a harmonic function of its vectorial argument, then $r^{-1}V(\mathbf{r}/r^2)$ is a harmonic function of $\mathbf{r}/r^2$. We need to emphasize the fact that the *geometrical reference frame must be the same* for the whole space, otherwise the formulation might be confusing. In a word, the theorem requires *the same reference frame*, and this does not seem to be obvious in all the presentations of the theorem. This is why we chose here a technical formulation, whereby the things must be kept proper, in view of their immediate application (Russell, 2013). The theorem allows us to characterize by harmonicity in a Euclidean background any given function: if it is harmonic on a certain space range, then its Kelvin transform is harmonic in the transformed range. As it happens, any component of the Newtonian force is harmonic at the place where the force acts. Therefore its Kelvin inversion components are also harmonic, but inside a sphere of inversion representing the star that generates the gravitational field.

As we already mentioned before, the sphere of inversion plays the part of a *gauging device* for a certain expanse of space, being therefore 'sanctioned', so to speak, to gauging that region. Mathematically speaking, the precise definition of such a device is inessential: one can simply assume that the radius of the gauging sphere is unity, and thus proceed to the general construction of the inversion with no problem. However, physically the things are quite complicate, at least for the decision on validity of the point approximation in estimating the position of a celestial body, if not for the process of interpretation in general. For instance, the Newtonian gravitation cannot be used in explaining the Kepler motion but in the limit of very small dimensions of Sun and planets with respect to the distance between them. At the cosmological level, the universe is not isotropic but at a certain scale, specifically that scale where the galactic nebulae can be considered material points with respect to the distance between them. We are therefore entitled to assume a certain homogeneity with respect to the dimensions of the bodies populating a region of space at a certain space scale, for which we accept a statistical characterization: they are a *normal population of spatially extended physical particles*. The statistics of space extension of the bodies of this population are then available by sampling (Nedeff, Lazăr, Agop, Eva, Ochiuz, Dumitriu, Vrăjitoriu & Popa, 2015), with the physical interpretation given by equations (4.31-37).

On the other hand, what is more important and thus necessary to be re-called here, is the fact that, as long as Madelung-type interpretation of the wave function is concerned, we can take this function as a physical instance of the definition of Ronald Fisher for the efficiency criterion. Indeed, the David Bohm's argument in its hydrodynamical variant, is referring to a quadratic relation between the wave function and the probability density. Here the compatibility between the nonstationary Schrödinger equation and the probabilistic character of the wave mechanics, is mediated by the very existence of the *Takabayasi tensor*, not by its integral which defines the Fisher information as we know it today. Therefore, if we limit ourselves to the local case, without going to make use of the integral, which is *the only operation imposing restrictions*, the Takabayasi tensor is essential: *everything in the space statistics depends on its existence*. This observation acquires a great authority if we take further notice of the fact that the description by Takabayasi tensor does not even require the normalisation of the density, which is essential in the definition of Fisher information. The Takabayasi tensor is also independent of any incidental optical notions that might be involved in the problem of transition between infrafinite and transfinite, this making it a necessary tool of the knowledge in general. Likewise the idea of hidden parameters connected with this tensors should also be a necessary tool in gauging the continuum.

In this differential context the hidden parameters are by no means 'hidden': *they are simply coordinates in the geometric space continuum*. In differential variant of statistics, therefore from an infrafinite point of view, the Takabayasi tensor — and through it the whole geometry of the hidden parameters — is physically fundamental. There must therefore be a circumstance objectively compelling us to accept the hidden parameters as natural coordinates in the three- dimensional phase space. In order to reveal this circumstance, we need to reconsider the idea of conformal metric in its utmost generality. The term 'utmost generality' carries here a little more precise meaning after all of the previous discussion: a *fragmentary defined surface of inversion*, as a system of diopters for instance, with physically defined 'fragments', and a general description of *the choice of origin of the reference frame.* These two items, even though related to the concept of reference frame, confer distinctive properties to the space itself, controlled by this reference frame: *it is a conformal space.* Then, as David Delphenich shows it, this space should however be endowed with *torsion* (Delphenich, 2002, 2013). In view of Élie Cartan's approach to geometry, we think that this is also the right way of approaching the physics. Let therefore get into some details of the subject.

Torsion Induced by Space Variations of Density

Recall that we are physically in the infrafinite space range, and we are to stay into that range. Then using equation (11.8) we can prove (Mazilu & Porumbreanu, 2018)

$$p_{\mathbf{x}}(\mathbf{x};\mathbf{s}) = \frac{\mathbf{m}^2}{\mathbf{x}^2}; \quad d\mathbf{m} \equiv \mathbf{\Psi} = p_{\mathbf{x}}(\mathbf{x};\mathbf{s})d\mathbf{x}$$

with $p_{\mathbf{x}}(\mathbf{x};\mathbf{s})$ given by equation (11.6). The metric $(d\mathbf{m})^2$ is therefore a conformal metric with respect to the Euclidean one $(d\mathbf{x})^2$, by a factor proportional to *the square of the probability density.* As up to this point the integrability condition is not yet of concern in the definition of the probability density, one can think that this situation is the one occuring in the transfer from a hypothetical empty space to the same space but filled with matter. Therefrom, a rule to which David Delphenich adheres in his theory: *the space filled with matter is conformal to the euclidean space, by a factor involving the density of matter.* He then relegates the differential situation we just described here to a deformation of *the physical coframe* of space. As long as by 'filling the space with matter' we understand an operation that does not change the scale of things mathematically — infrafinite remains infrafinite, finite remains finite, etc. — we think that Delphenich's approach is the only sound way to physics. Indeed, if we can express the metric of space in the conformal euclidean form:

$$\mathbf{g} = \rho(\mathbf{x}, t) \cdot \mathbf{1} \quad \therefore \quad \mathbf{g} = \mathbf{\Psi} \otimes \mathbf{\Psi} \qquad (11.22)$$

where $\mathbf{\Psi}$ are the coframe vectors having some differential forms as components:

$$\Psi^i(\mathbf{x}, t) = \Psi^i_k(\mathbf{x}, t)dx^k. \qquad (11.23)$$

Obviously, in the usual manner, we need to put the components of $\mathbf{\Psi}$ in connection with the square root of the density of matter. Delphenich takes note that *in a theory of distant parallelism,* this would be only a special case, very particular indeed. However, still following this line of thought, and defining the components (11.23) by equation

$$\Psi^i_k(\mathbf{x}, t) = \sqrt{\rho(\mathbf{x}, t)} \cdot \delta^i_k. \tag{11.24}$$

Delphenich concludes that the *square root of density* becomes 'more fundamental than the density itself'. Starting from this, another important observation of Delphenich relates to the fact that the deformation is, in general, *not an usual diffeomorphism*, even in the classical case when, as a rule, it is modeled by a diffeomorphism: in fact the model must always be mediated by some sound physical considerations.

In the present context, the differential forms from (11.23) should first be exact differentials, which entails the condition:

$$d \wedge \Psi^i(\mathbf{x}, t) = 0 \quad \therefore \quad \partial_j \Psi^i_k(\mathbf{x}, t) = \partial_k \Psi^i_j(\mathbf{x}, t). \tag{11.25}$$

However, it is to be noticed that (11.24) *does not satisfy this condition*. More precisely we have:

$$d \wedge \Psi^i(\mathbf{x}, t) = \Omega \wedge \Psi(\mathbf{x}, t), \quad \Omega \equiv \frac{1}{2} dx^k \frac{\partial(\ln\rho)}{\partial x^k}. \tag{11.26}$$

At this point Delphenich noticed that *the symmetric differential* of the 1-form Ω,

$$d\Omega \equiv \frac{1}{2} \frac{\partial^2(\ln\rho)}{\partial x^i \partial x^j} dx^i dx^j \tag{11.27}$$

defines the Takabayasi tensor:

$$\tau_{ij} dx^i dx^j \equiv \frac{\hbar}{2m} \rho \cdot d\Omega. \tag{11.28}$$

Then the equation (11.26) shows that the *Takabayasi tensor is related to the torsion of space*. More precisely, even if the equation (11.25) is not satisfied, the space still possesses *a connection*, given by the coefficients

$$\Gamma^i_{jk}(\mathbf{x}, t) = \partial_j \Psi^i_k(\mathbf{x}, t) \tag{11.29}$$

which splits, by symmetrization in the lower indices, into two parts defining the *symmetric connection* and the *torsion tensor*, respectively:

$$\begin{aligned} 2\overline{\Gamma}^i_{jk}(\mathbf{x}, t) &= \partial_j \Psi^i_k(\mathbf{x}, t) + \partial_k \Psi^i_j(\mathbf{x}, t); \\ 2S^i_{jk}(\mathbf{x}, t) &= \partial_j \Psi^i_k(\mathbf{x}, t) - \partial_k \Psi^i_j(\mathbf{x}, t). \end{aligned} \tag{11.30}$$

This shows that such a metric theory describes a nonholonomic mechanics, whereby the torsion turns into a measure of the nonholonomy. Now, having in view the special situation given by (11.24), both the symmetric connection and the torsion are determined solely by the density:

$$2\overline{\Gamma}^i_{jk}(\mathbf{x}, t) = \delta^i_k \partial_j \sqrt{\rho(\mathbf{x}, t)} + \delta^i_j \partial_k \sqrt{\rho(\mathbf{x}, t)};$$
$$2S^i_{jk}(\mathbf{x}, t) = \delta^i_k \partial_j \sqrt{\rho(\mathbf{x}, t)} - \delta^i_j \partial_k \sqrt{\rho(\mathbf{x}, t)}. \tag{11.31}$$

When calculating the *torsion covector*, we then have:

$$S_j(\mathbf{x}, t) \equiv S^i_{ji}(\mathbf{x}, t) = \partial_j \sqrt{\rho(\mathbf{x}, t)} \tag{11.32}$$

which proves our conclusion. In other words, the torsion covector should be given by the gradient of the density of matter.

The Reference Frame and the Torsion

The previous conclusion of David Delphenich is instrumental: it gives a unique way to physics whereby the presence of matter can be described in simple words. Namely, as equation (11.22) shows, if the *space of residence of matter* is Euclidean, then the presence of matter is manifest by the fact that the metric of space becomes conformal, with the conformity factor given by the density of matter. Two critical points arise here. First, we do not know anything about the space of residence of matter, forasmuch as it does not get by anything to our senses. All we know for sure is that a space of the residence of the matter of zero density can be characterized as Euclidean. On the other hand, the way in which the torsion is introduced here − see equation (11.26) − suggests a non-Euclidean geometry. Delphenich himself, follows this line. However, as nothing is sure about the residence space of the matter, we can limit ourselves to the Euclidean space, and the problem occurs if such a bacground space is compatible with the idea of torsion. This idea motivated a critical review of Élie Cartan with the occasion of a problem of the same nature as the present one: the unitary theory of Albert Einstein, based on distant parallelism. First, here is an excerpt from Cartan, which we chose for illustrating the general case here, which will be then discussed based on it:

> ... It is easy to realize the most general way to define an *absolute parallelism* in a given Riemannian space. Attach, indeed, to the different points of this space *rectangular reference systems*, or *frames*, and this according to an *arbitrary law*; it is then sufficient *to agree* that two vectors of any origins A and B *are parallel*, or better equipollents, *if they have*

the same projections along the axes of reference systems of origins A and B; *these reference systems will be then parallel themselves.* There are, therefore, in a given Riemannian space, an infinity of possible absolute parallelisms, for *the law according to which one attaches a rectangular frame to a point in space is completely arbitrary;* however, we must notice that if all the rectangular frames are rotated in the same way around their origins, one gets the same absolute parallelism; therefore, *one can define once and for all the frame attached to a particular point in space.* [(Cartan, 1931); *our translation and Italics; see also* (Delphenich, 2011), pp. 202 - 211]

Therefore, attaching a reference frame is, according to Cartan, *a matter of gauging* in the acceptance of Nicholas Georgescu-Roegen for defining the 'finite': 'define once and for all the frame attached to a particular point'. However, it is to be noticed that this frame is taken by Cartan as 'rectangular', and it is not hard to conclude that this condition is only taken here in order to avoid further arbitrariness: such an assumption relegates the gauging only to the orthogonal group. This kind of reference frame epitomizes the classical 'box locating' of a position, which seems to be currently contemplated almost exclusively, especially in the intuitive imagery of a reference system. In spite of this particular choice of the frame, the Cartan definition of the absolute parallelism still remains the most general one, especially by comparison to its definition by continuity (Levi-Civita, 1916) and, what is more important, it is the only one closer to a physical spirit of definition, particularly when it comes to theoretical statistical or stochastic processes calculations. For once, the case appears to be one of a kind, both among the ideas of Élie Cartan himself and those of the differential geometry in general. As far as we are aware, this idea of definition of parallelism cannot be found, either in his previous works, or in the works following the one just cited, at least not in such an unequivocal form of expression, to say nothing of the geometrical works of others.

Indeed, one can say that the previous excerpt defines the absolute parallelism by a 'mnemonic scheme': the components of a vector, 'recorded' somehow, are reproduced by orthogonal projections in each and every one of the frames attached to positions in space according to an arbitrary, but nevertheless specified, rule. In such a situation, one can say that the frames are also parallel. We find this approach to geometry closer to the spirit of modern physics, forasmuch as, first of all, it contains the suggestion that the definition of the frame parallelism in a given space filled with matter *depends on the possibility of recording and transmitting the information* within that space. Secondly, Cartan's definition admits an important 'reciprocal': *we can define a class of parallel frames, once we have at our disposal three numbers physically representing the components of a vector.* We will offer shortly an important example concerning the most important mechanism of transmitting information in space: *the propagation of light* — of a signal in general — which is the universal carrier of information in the known universe.

Meanwhile let us go a little deeper into this manner of building the geometry, by showing that it is genuinely related to the *definition of torsion.* On this unique occasion, Élie Cartan insists upon feasibility of what, following his wording, we like to call an 'Euclidean mentality' which obviously leads to abandoning the idea of curvature, as its name would imply, but brings instead the torsion to the fore. According to Cartan, the torsion is contained in some kind of indecision of the vector representation in a Riemanian space and that in an entirely natural manner, as far as the Euclidean mentality is involved. Quoting:

> It is known that in the usual geometry the coordinates of a point M, referred to a system of rectangular axes of origin O, are the projections of the vector $\overrightarrow{OM}$ along these axes; we can still get them by joining O and M with a broken line, and then taking the sum of the projections of different parts of this line. One can even take a curved line, considered as a limit of a broken line. Now, imagine an observer located in a Riemannian space with absolute parallelism, having however an Euclidean mentality. If this observer, placed in O and adopting a system of rectangular axes of origin O, *wants to calculate the coordinates which he must assign to a point* M (*our Italics*), he will join O with M by a continuous line, and will proceed as we just have shown: he will consider the line OM as a geometric sum of a very large number of minute vectors; he will transport them in O parallel with themselves, and then will take their geometric sum: thus he will find a vector of origin O, which he will consider as equipollent to the line OM, and whose *projections upon axes shall be the coordinates he sought for* (*our Italics*). If the observer joins O and M by another line, he will be led to consider it as equipollent to a second vector, which *generally will not be the same with the first vector.* In other words, the different lines joining O and M are not all equipollent to the same vector.
>
> The issues can be presented yet another way. If in the Euclidean geometry one considers a closed contour, or cycle C, pursued in a certain direction, it is equipollent to a null vector, according to a fundamental theorem of the vector calculus; in a Riemannian space with absolute parallelism this is no more the case: the cycle C is equipollent to a

certain vector, which we shall call *torsion vector*. Only in the Euclidean space we will have, for all cycles, null torsion vector. [*Our translation; Italics in the original, except as indicated; see also* (Delphenich, 2011), *loc. cit.*]

As an observation connected to this definition of the torsion, we now describe, as promised above, the concept of matter that can emerge from a quotidian example: that of the Earth as a planet (Mazilu & Porumbreanu, 2018). First, let us apply the above idea of Élie Cartan to an operational definition of the vertical direction on Earth surface accessible to our regular displacements. This, of course, will get us some unit vectors to be represented as points on the unit sphere. With any three of these directions, we can construct an estimator of the position of center of Earth, viz. the point toward which the the force of weight of earthly bodies presumably acts. This center is, nonetheless, never unique, but varies within a region inside the Earth, to which we never have access, for the space itself has never access there. According to Cartan's idea this region has a finite extension which can be given by a torsion vector. In general, the matter of Earth's nucleus — or even more generally, of a spatially extended particle — is characterized by the torsion vector. In physics, as we have shown above, this idea made its way only in our times (Delphenich, 2013), and it is related to a concept of density able to properly generalize the Newtonian concept in order to allow for a SRT.

The Torsion and the Waves

Speaking of the Earth and related sciences, it is just a matter of course to come to earthquakes and seismograms. Fact is that the position assigned to the recording place of a seismogram reproduces the Cartan's definition of a position from the long excerpt above, obviously with some proper choice of alternatives. One of these — the most important one in fact — would be that, in the case of a seism, the procedure of relying on joining the points in space, which, as presented by the great geometer may appear to some as subjective, or at least as inessential, is practically replaced by *the signal propagation* from the source to seismograph. For once, the knowledge of an equation of propagation of the signal is therefore mandatory for physically tackling the problem of positioning: the different Cartanian lines joining the position of the source of seism with the position of seismograph, are simply paths followed by the perturbation representing the signal, and thus they should be physically defined.

In such a process of 'objectification', if we may, we meet some well-known concepts of physics. For instance, a Cartanian line joining two positions between which the perturbation propagates in a material medium, is to be interpreted as a *mathematical ray* describing the propagation. The procedure of construction of a physical ray in geophysics even carries a specific name in technical terms: *ray tracing*. Obviously, between the source and the recording site, there are a multitude of mathematical rays, and these are contained in a volume reminding us the coherence volume from the physical optics, thus defining a *physical ray*. In seismology this volume is known for a while as the *Fresnel volume*, the name reminding its origins in the physical theory of light [see (Červený, 2005); especially § 3.1.6]. This volume depends on the frequency recorded at the seismograph, and obviously its definition should contain this frequency in a certain way. In the simplest of the cases, the original case of defining a Fresnel volume in fact, the definition involves a modification of the Fermat's principle, [(Kravtsov & Orlov, 1982); (Kravtsov, 1988)]. However, the suggestion immediately presenting itself according to Cartan's philosophy illustrated in the excerpt above, is that the *Fresnel volume is inherently connected to the torsion of the space* containing the matter within which the signal is propagating. The existence of this very volume is therefore a necessary physical condition for the possibility of describing the propagation. For the case in point, it should then be formally correlated with the *deformation of the surface seismically delimiting the Earth*.

Fact is that, in localizing the source of an earthquake, we do not have at our disposal but the signal recorded in a place on the inhabited surface of the Earth. The record itself is in the form of *an apparently periodic perturbation, with an amplitude variable in a time sequence locally devised*: this is the seismogram. Physically speaking, we can say that such a record reproduces the deformations of the Earth surface, by a succession of relative positions; again, this succession is defined by means of a locally measured time, with an arbitrary clock. It is therefore quite natural to think that such a succession of local positions is a direct consequence of the perturbation propagation from the interior of the Earth to surface, so that the quake itself can be globally envisioned as a consequence of this propagation from its source to the surface. Geophysics — the science in charge with the solutions of the mathematical problems posed by the earthquakes — approaches the problem by simplifying it to essentials. In the most common of simplifications, we have a propagation of an undulatory signal through an ideally homogeneous material, viz. *through matter*, as we conceive it here. That 'most common' simplification involves a differential equation describing the propagation, usually the D'Alembert equation:

$$\Delta u = \frac{1}{c^2} \frac{\partial^2 u}{\partial t^2}. \tag{11.33}$$

Here 'u' is the instantaneous elongation of the signal, function of position and time in the volume occupied by matter, and 'c' is the velocity of propagation of the signal, depending on the elastic properties of the matter.

The mathematical problem comes now down to *constructing a correlation between different positions from the space occupied by matter, at different times.*

Classically speaking, for this we need a reference frame whereby the position is located, and a clock to mark the moments of time, in the ways that these two operations are usually thought to be accomplished. Specifically, the reference frame is usually Euclidean, and *the representation of the coordinates as lengths along any three reciprocally orthogonal directions* is just implicit, so to speak. The correlation must be constructed based upon the equation (11.33), which is actually considered as its infinitesimal form. In these circumstances G. L. Shpilker took notice of the fact that a real seismogram complicates the things quite unexpectedly (Shpilker, 1982): *the position of the point of recording must be defined not by lengths, but by three numbers having certain algebraical properties, necessary in order to comply with the definition of the recorded signal!* Taken as the components of a vector, these three numbers actually define a class of parallel reference frames in the sense of Cartan. Some specific explanations are in order here.

Shpilker's theory starts from the observation that a seismogram can never be represented by the simple harmonic oscillation which, in the geometry of the equation (11.33), would represent locally a plane wave. A general form of the recorded signal, having any realistic appearance at all, would be as a complex-valued function of a locally devised time sequence that serves to ordering the elongations recorded in the seismogram. Such a realistic appearance would be given, in Shpilker's representation, through a function of time having the form:

$$v_1 = |A|e^{i\alpha_0}e^{(i\omega+\beta)(t-t_0)} \tag{11.34}$$

with A, t_0, α_0, b and ω − five constants, to be extracted from the seismogram itself. The general observation is that *the form of this signal as a function of time* is all we are able to know with a certain degree of confidence: the rest, starting from the very idea of propagation, the equation representing it etc., is just a series of hypotheses. However, we need to emphasize from the very beginning, and subsequently certify, that the philosophy beyond this procedure is universal. Let us expound a little on this statement.

First, the equation (11.33) − or any equation of propagation for that matter − incorporates only a part of the hypotheses. Admitting, in order to not complicate things in settling our ideas, that the propagation is described by D'Alembert equation as given in (11.33), the problem of correlating two points in space is usually solved in physics by the Green function corresponding to this equation. However, within Shpilker's philosophy the emphasis is significantly changed: *it falls upon the correlation of the recorded signal with the equation of propagation*, which is a step of knowledge generally bypassed in the regular usage of an equation of propagation. For, it is quite clear that the equation (11.34), a *product of experiment* in the Cartan's order of things geometrical, bestows a *physical content upon the space position where the signal is recorded*. And this physical content is described by (11.34) through the intermediary of a local time sequence, in exactly the same manner in which a uniform motion of a classical material point, for instance, bestows a physical content to a certain time sequence, obtained with an arbitrary clock. Let us analyze the way in which, according to G. L. Shpilker, such a physical content should be brought to bear on the geometry.

Once we have at our disposal the equations (11.33) and (11.34) — in general, as we said, an *equation of propagation and a physical content of a local sequence of time* — the Shpilker's argument follows quite a simple logic, customary we might say: one must accept that any recorded signal is a solution of the D'Alembert equation, for *this equation defines the concept of signal within matter*, and the recorded signal itself is, obviously, such a signal. The only condition is that the *surface of earthquakes should be a matter surface*, a quality that makes out of it a surface of separation of the matter from space. Notice now that the representation (11.34) of the recorded signal *does not contain explicitly any space position*, be it that of the source of the seism or of the position of seismograph, but just some parameters to be read on the seismogram. It would be therefore virtually impossible to set this physical content of the signal in connection with the equation of propagation (11.33), if one does not assume that the position of the recording point in space is somehow contained implicitly among the parameters representing this physical content: α_0, ω, β. Shpilker writes the solution of the equation (11.33) in the form:

$$u(\mathbf{x}, t) = A e^{\langle \xi | x - y \rangle + c \sqrt{\langle \xi | \xi \rangle} (t - t_0)} \qquad (11.35)$$

where ξ is an arbitrary complex vector. Obviously, this solution satisfies D'Alembert equation, both in the variables $(\mathbf{x}, t)$, and in the variables $(\mathbf{y}, t_0)$, no matter of the vector ξ and amplitude A, which is here a complex number:

$$A = |A| e^{i \alpha_0}. \qquad (11.36)$$

Consequently, the equation (11.35) epitomizes *a correlator analogous to the classical Green function*, of two 'legal signals', the legality being defined according to the criterion of definition of *admissible signals* by D'Alembert's equation: theoretically this signal must be found all over the places within matter, therefore both at the location of emission and the location of recording. Except that now the functional form of the signal at the emission position, or in fact during propagation, is somewhat more realistic, inasmuch as it is not *a priori* defined, but empirically, with a physical content defined *in the manner we define the recorded signal*.

Now the solution of the problem comes down to matching this theoretical representation with the recorded signal from equation (11.34). In order to do this, Shpilker uses the freedom offered by the arbitrariness of the vector ξ :

in the surface delimiting the Earth seismically — the *surface of quakes*, as we would like to call it — he takes the signal as being of the form

$$v(\mathbf{x}, t) = A e^{\langle z \otimes (k+il) | x-y \rangle + (i\omega + \beta)(t - t_0)}. \qquad (11.37)$$

This signal reduces to that from equation (11.34) for $|x - y\rangle = |0\rangle$, which means that $|y\rangle$ *may be taken as the position of the point of recording*. Then, again, the function (11.37) is a solution of the D'Alembert equation, this time, however, in a special conditions whereby over the recorded signal one overlays another signal, which needs to be conveniently described in order to account for the conditions in which the measurement is performed.

Before any explanation on these conditions, a word about the notations from equation (11.37): the vector $\mathbf{z}$, as well as ξ for that matter, is unknown. The vector $\mathbf{k} + i\mathbf{l}$ is an arbitrary complex vector, submitted by Shpilker to the constraints:

$$k_1^2 + l_1^2 = k_2^2 + l_2^2 = k_3^2 + l_3^2 = \tau^2 \qquad (11.38)$$

where τ is an arbitrary real number. Further on, one denotes

$$|z \otimes (k + il)\rangle \equiv \begin{pmatrix} z_1(k_1 + il_1) \\ z_2(k_2 + il_2) \\ z_3(k_3 + il_3) \end{pmatrix} \qquad (11.39)$$

so that this is just *a vector with complex components given by the diagonal entries of the complex matrix* $\mathbf{z} \otimes (\mathbf{k} + i\mathbf{l}) \equiv |z\rangle\langle k + il|$.

Now, coming back on the track of our discussion, Shpilker claims that in order to get a correct 'reconstruction' of the field from the recorded signal as defined by equation (11.34), the coordinates of the position of recording must be expressed by the ratios:

$$y_i = \frac{l_i}{k_i}; \quad i = 1, 2, 3 \qquad (11.40)$$

submitted to the conditions

$$y_2 \neq y_3; \quad y_3 \neq y_1; \quad y_1 \neq y_2 \qquad (11.41)$$

which are thus necessary and sufficient for a reconstruction of the field from recorded data. Therefore Shpilker's local coordinates *are not regular coordinates*. According to Cartan's point of view, it is more proper to say that *they define in fact a reference frame*: that reference frame in which the coordinates of the position of recording point are given by the vector $\mathbf{y}$ having the components (11.40), submitted to the conditions from equation (11.41).

Let us show that in order to have a solution of the problem in the form from equation (11.37), the Shpilker's demands have to be met indeed. In order to do this, notice that from the equations (11.34) and (11.35) one gets:

$$u(\mathbf{y}, t) = v_1 \quad \therefore \quad \begin{cases} c^2(\xi_1^2 + \xi_2^2 + \xi_3^2) = \beta^2 - \omega^2 + 2i\beta\omega; \\ c^2(|\xi_1|^2 + |\xi_2|^2 + |\xi_3|^2) = \beta^2 + \omega^2, \end{cases} \quad (11.42)$$

by the virtue of the fact that ξ has complex components in general. In fact, according to equation (11.37) these components are defined by equation (11.39), so that the arbitrariness of the vector ξ, having six real components, is transferred into the ambiguity of the vectors $\mathbf{z}$, $\mathbf{k}$ and $\mathbf{l}$, which involve nine real components. The situation becomes normal if we have three relations connecting these last two vectors, which must be *measurement constraints*, as those given by Shpilker in equation (11.38), which can be taken just naturally as such. Therefore, using the equations (11.40) in (11.42), the components of vector $\mathbf{z}$ can be found as solutions of the linear system:

$$c^2[(k_1^2 - l_1^2)z_1^2 + (k_2^2 - l_2^2)z_2^2 + (k_3^2 - l_3^2)z_3^2] = \beta^2 - \omega^2;$$
$$c^2[k_1 l_1 z_1^2 + k_2 l_2 z_2^2 + k_3 l_3 z_3^2] = \beta\omega;$$
$$c^2[(k_1^2 + l_1^2)z_1^2 + (k_2^2 + l_2^2)z_2^2 + (k_3^2 + l_3^2)z_3^2] = \beta^2 + \omega^2,$$

which is obviously equivalent to the system:

$$c^2[k_1^2 z_1^2 + k_2^2 z_2^2 + k_3^2 z_3^2] = \beta^2;$$
$$c^2[k_1 l_1 z_1^2 + k_2 l_2 z_2^2 + k_3 l_3 z_3^2] = \beta\omega; \quad (11.43)$$
$$c^2[l_1^2 z_1^2 + l_2^2 z_2^2 + l_3^2 z_3^2] = \omega^2.$$

This system is in turn compatible, and has unique solution if, and only if, its principal matrix is nonsingular. The determinant of this matrix can be easily calculated, and gives:

$$\begin{vmatrix} k_1^2 & k_2^2 & k_3^2 \\ k_1 l_1 & k_2 l_2 & k_3 l_3 \\ l_1^1 & l_2^2 & l_3^2 \end{vmatrix} = (k_2 l_3 - k_3 l_2)(k_3 l_1 - k_1 l_3)(k_1 l_2 - k_2 l_1).$$

Therefore the compatibility of the system (11.43) comes to the fact that none of the projections of the real vector $\mathbf{k}$ on the planes of coordinate should be collinear with the corresponding projection of the real vector $\mathbf{l}$. Solving the system (11.43) results in

$$z_1^2 = \frac{1}{c^2} \frac{l_2 l_3 \beta^2 - (k_2 l_3 + k_3 l_2)\beta\omega + k_2 k_3 \omega^2}{(k_3 l_1 - k_1 l_3)(k_1 l_2 - k_2 l_1)}$$

and its even permutations over indices 1, 2, 3. Considering now the definitions from equation (11.40), we will have right away:

$$z_1^2 = \frac{1}{c^2 k_1^2} \frac{y_2 y_3 \beta^2 - (y_2 + y_3)\beta\omega + \omega^2}{(y_3 - y_1)(y_1 - y_2)} \qquad (11.44)$$

and two more, given by the circular permutations over indices 1,2,3. This proves the necessity and sufficiency of the conditions (11.41) of Shpilker, showing moreover that the vector **k** must have all its components nonnull for a reconstruction of the field in finite terms. Consequently, the triple $|y\rangle$ represents here the position of the recording point of the earthquake, according to its definition contained in the particular relation between D'Alembert equation – describing the signal propagation – and the functional form of the recorded signal.

For G. L. Shpilker – as well as for the whole geophysicists' community, in fact – such a resolution of the problem of quakes is essential. Indeed, the seismogram is actually a singular expresssion of a limiting condition in space and time for an equation of propagation – for the case in point the D'Alembert equation (11.33) – and for instance a conceivable Cauchy problem of this case cannot be solved with a boundary condition in a single point. Usually, for solving such a problem one would need conditions over *the entire surface of the Earth, defined by the existence of quakes.* First of all, such a surface cannot be defined itself, even if we disregard the idea of seismogram, to say nothing of the fact that one cannot place seismographs all over the places where an earthquake is felt, in order to make the necessary measurements. It is therefore instrumental, indeed, to built a signal as the solution of the equation of propagation, *starting from data recorded sporadically,* insular data at best. Which is what Shpilker's theory accomplishes in a brilliant way. This approach has, however, much more general connotations, even fundamental we should say, from the point of view of the theoretical physics. They can be extracted even limiting our considerations to the classical differential geometrical idea of adaptation of a reference frame to a surface embedded in space, and the description of the deformation of such a frame (Delphenich, 2013). However, with the idea of torsion in constructing a physical model of matter, we are heading towards other, more fundamental realm.

Chapter 12. Theory of Nikolai Alexandrovich Chernikov

One can justly say that the unitary theory of Albert Einstein, based on the concept of absolute or distant parallelism [see (Einstein, 1930); see also (Delphenich, 2011) for English translations, a comprehensive bibliography and pertinent commentaries on this subject] represents still another logical attempt — among so many others explored by the great physicist along time — to consolidate the initial idea of theory of relativity, according to which the metric of the world is determined by matter. Einstein must have had realized, as he did in many other cases transparent in the various interventions to amend the theory — see (Goenner, 2004, 2014) and (Renn, 2007), for details, critical discussions and a comprehensive bibliography — that by proposing the metric tensor as unique representative of the matter, he drifted apart from the initial geometrical principle, which states that the matter establishes in fact the *space connection*, as suggested by the Newtonian theory. Fact is that the space connection can be defined quite independently of the metric. Being nonetheless impossible to give up the idea of metric, Einstein may have searched for a way in which the metric tensor should be correlated with the connection of space, and such a way ensues from the very manner in which the natural connection of a Riemannian metric is calculated (Misner, Thorne & Wheeler, 1973). As a consequence of his idea of *distant parallelism* — or *absolute parallelism*; the notion is best described by lie Cartan, in a portrayal that, as indicated before, we would like to term as 'informational' (Cartan, 1931) — Einstein proposes the *equations of compatibility between metric and connection* and not the metric *per se*, as being theoretically essential. In this case the metric tensor is no more constrained to be a symmetric matrix. In modern terms one can say that the matrix symmetry expressing the invariance of the quadratic metric itself when an arbitrary skew-symmetric part is added to the metric tensor, is broken if the fundamental equations of the theory are the compatibility equations between metric and connection. One can further say that the tensor from which the metric is calculated can be no more just a metric tensor, but a general tensor that should be properly called a *fundamental tensor*.

Just about the same time with the modern beginnings of the theories of distant parallelism, and from exactly the same general philosophical reasons, Max Born and Leopold Infeld constructed the celebrated model of the nonlinear electrodynamics, with the declared task of avoiding the singularities brought in electrodynamics by the classical concept of material point (Born & Infeld, 1934). The Born-Infeld electrodynamics accomplishes, in specific details, the Hilbert's idea from 1915 regarding the very foundations of physics [for a critical discussion and even an English translation of Hilbert's works on the subject see (Renn, 2007), Volume 4]. The essential theme of Hilbert's idea was a variational principle using in its formulation the Riemannian volume element in building an invariant Lagrangian density. Max Born and Leopold Infeld have then shown that in the case of the space-time continuum of the special relativity this principle is equivalent with the invariance of the *non-Riemannian volume*, based on a *general fundamental tensor*. The generality is here understood in the sense that this tensor *has no matrix symmetry*: it has a symmetrical matrix part which alone is used in calculating the metric. On the other hand, the skew-symmetric part of the fundamental tensor has properties which, in specific conditions, are formally identical with those of the electromagnetic field. Those conditions are indeed characteristic for the classical definition of such a field.

Enters Chernikov

The essential point – in our opinion! – of the underlying mathematical theory of both Einstein's ideas and Born-Infeld electrodynamics was made obvious by Nikolai Alexandrovich Chernikov, and will be rendered in a specific detail in what follows, along with the consequences of the unitary theory which we hold as fundamental for the knowledge in general. That point can be briefly summarized by the statement that the Born-Infeld equations of the nonlinear electrodynamics *are valid in any dimension, in a geometry based on Einstein compatibility equations*, whereby they physically signify the *vanishing of the torsion covector*. Perhaps this is the best time to explain why do we insist on a theory valid 'in any dimension'. Actually, what we have in mind is a three-dimensional theory.

Fact is that the matter 'fills' the space of our intuition, which is three-dimensional. This 'filling' is irregular in most cases, as Newton himself noticed in *Principia*, being represented by physical structures, accessible to our senses. These are simply *matter penetrated by space*, as the physics of the last two centuries or so tells us. The four-dimensional universe of relativity, however, tells another story related, as it were, to the *time concept*. And, as Einstein explained in his celebrated article on foundations of special relativity (Einstein, 1905), this side of the time concept involves, even though not exclusively, the electrodynamics. For, the time, once it gets out of our intuition, becomes a concept having two distinct differentiae, sometimes mixed in physics uncontrollably with one another. On one hand, the time has the classical property of being *a continuity parameter*. This property is extracted from the perception of motion, and surfaced in relativity through the idea of geodesics, which further led to noticing the importance of a connection of space. Electrodynamics, on the other hand, requires the time in another instance: *as a parameter ordering a sequence of events*. This is the essential content of the Einstein 1905 paper, and the issue gets through all of the ulterior developments of the intellect of the great theoretician, to the extent in which they are accessible to us, of course. It may help to make the concept of time a little more comprehensible along this line, if we notice that Kurt Gödel's works from 1949, hesitantly sanctioned by Einstein himself [see (Gödel, 1949, 1952); (Einstein, 1949)], are the first works in which the two differentiae of the time concept came blatantly at odds, shaking our intuition with the idea of travelling in the past. For, the result of Gödel's work would mean, in turn, possibility of influencing the past, and this idea nourished a lot of popular speculations, inappropriately promoted, as usual, by media for entertaining. The scientific consensus here seems to be that a geodesic motion in a Gödel universe is practically impossible, but even this opened another line of speculations. Be it as it may, this attitude of physics with respect to the concept of time, is a warning for properly distinguishing among the concepts it forced upon our intellect. And the basic such concept is, of course, that of the matter itself: it cannot fill but a three-dimensional space, and the manner of filling leads to a Riemannian manifold. Further on, the presence of matter leads to the consideration of torsion for completing the connection of space filled with matter, but the dimension of this space is still three. Therefore we need to describe this situation, and then, based on this description, we can go further on and describe a kinematics, or even a dynamics of matter.

Let us therefore show how the torsion can be related to the fundamental tensor, using the calculation line of Chernikov himself. The work we follow closely is that concerning a metric space of general dimension, even or odd, endowed with a fundamental tensor field determining the metric by its symmetric part (Chernikov, 1977). This work can be considered as a sort of synthesis of the results of some previous works of the same author (Chernikov, 1976). We think that these works are sufficiently important in order to deserve a representation like the one following here (and, perhaps, an even better representation!), somewhat detailed up only to a point, of the ideas they promote, and especially of their way of mathematical realization. These ideas scarcely occurred in the specialty literature at large, and therefore they were not exposed to any critical consideration from the part of scientific community.

The compatibility equations between metric and connection proposed by Einstein are given by vanishing of the covariant derivative of the fundamental tensor, when this one is constructed with the connection coefficients:

$$\frac{\partial g_{\alpha\beta}}{\partial x^\gamma} - g_{\alpha\sigma}\Gamma^\sigma_{\gamma\beta} - \Gamma^\sigma_{\alpha\gamma}g_{\sigma\beta} = 0. \tag{12.1}$$

As we just mentioned, when based on these equations the theory does not necessarily require the symmetry of the metric tensor. We can therefore assume the existence of a fundamental tensor **g**, with no special matrix symmetry, and which can thus be written in the general form:

$$\mathbf{g} = \mathbf{h} + \mathbf{a}. \tag{12.2}$$

Here **h** is the symmetric part of the fundamental tensor, which alone determines the metric, for we necessarily have:

$$(ds)^2 \equiv g_{\alpha\beta}dx^\alpha dx^\beta = h_{\alpha\beta}dx^\alpha dx^\beta.$$

Further on, **a** is the skew-symmetric part of the fundamental tensor, guaranteeing the validity of this equation of quadratic metric, and remaining to be interpreted from a physical point of view. From equation (12.2) the two matrix components of the fundamental tensor can be calculated by the obvious relations:

$$2h_{\alpha\beta} = g_{\alpha\beta} + g_{\beta\alpha}; \quad 2a_{\alpha\beta} = g_{\alpha\beta} - g_{\beta\alpha}.$$

As the connection itself is not expected to be symmetric in the conditions of equation (12.1), one needs to try a decomposition of the connection symbols similar to that of the fundamental tensor:

$$2\overline{\Gamma}^\mu_{\alpha\beta} = \Gamma^\mu_{\alpha\beta} + \Gamma^\mu_{\beta\alpha}; \quad 2S^\mu_{\alpha\beta} = \Gamma^\mu_{\alpha\beta} - \Gamma^\mu_{\beta\alpha}. \tag{12.3}$$

These notations define here the *mean connection* and the *torsion* respectively, which is a mixed tensor with two indices of covariance and one index of contravariance (Vrânceanu, 1962). The connection in general is not a tensor. Denoting here the covariant derivatives by symbol ∇, the covariant derivative in the mean connection of the fundamental tensor should be

$$\overline{\nabla}_\gamma g_{\alpha\beta} = \frac{\partial g_{\alpha\beta}}{\partial x^\gamma} - g_{\alpha\sigma}\overline{\Gamma}^\sigma_{\gamma\beta} - \overline{\Gamma}^\sigma_{\alpha\gamma}g_{\sigma\beta}. \tag{12.4}$$

This means that the equation (12.1) can be written as

$$\overline{\nabla}_\gamma g_{\alpha\beta} = g_{\alpha\sigma}S^\sigma_{\gamma\beta} + S^\sigma_{\alpha\gamma}g_{\alpha\beta}. \tag{12.5}$$

Therefore, the covariant derivative of the fundamental tensor in the mean connection is determined by the tensor itself and by the torsion tensor.

211

Now, in general, the fundamental tensor can even be singular. However, in the case it is nonsingular, we can define its reciprocal, denoted $\tilde{g}$ having the entries $\tilde{g}^{\alpha\beta}$, say. Indeed, by the rule of calculation of the determinant based on a line or a column, we have

$$g_{\alpha\mu}m^{\alpha\nu} = g\delta_\mu^\nu = g_{\mu\beta}m^{\nu\beta} \tag{12.6}$$

where $\mathbf{m}$ here is the matrix of minors of $\mathbf{g}$, and 'g' means the determinant of $\mathbf{g}$. This relation is independent of the value of determinant: the matrix can be singular too, the relation still works. However, in cases when $g \neq 0$, and therefore if the matrix $\mathbf{g}$ is nonsingular, we can define the reciprocal by $\tilde{\mathbf{g}} \equiv \mathbf{m}/g$, so that the equation (12.6) can be written in the usual form:

$$\tilde{\mathbf{g}}^t \cdot \mathbf{g} = \tilde{\mathbf{g}} \cdot \mathbf{g}^t = \mathbf{1}. \tag{12.7}$$

Here the upper 't' index means the transposed matrix, and $\tilde{\mathbf{g}}^t$ obviously means the inverse of $\mathbf{g}$ in g in the usual matrix multiplication, and $\mathbf{1}$ denotes the identity matrix, as usual. If, by analogy with equation (12.2), we decompose $\tilde{\mathbf{g}}$ into its symmetric and skew-symmetric parts, i.e. $\tilde{\mathbf{g}} \equiv \tilde{\mathbf{h}} + \tilde{\mathbf{a}}$, then the condition (12.7) is equivalent with the following two relations:

$$\mathbf{a}\tilde{\mathbf{h}} = \mathbf{h}\tilde{\mathbf{a}}; \quad \tilde{\mathbf{h}}\mathbf{a} = \tilde{\mathbf{a}}\mathbf{h}. \tag{12.8}$$

On the other hand, in terms of the reciprocal tensor, the equations of compatibility (12.1) are

$$\frac{\partial \tilde{g}^{\mu\nu}}{\partial x^\gamma} + \tilde{g}^{\alpha\nu}\Gamma^\mu_{\alpha\gamma} + \tilde{g}^{\mu\alpha}\Gamma^\nu_{\gamma\alpha} = 0$$

so that for the covariant derivative in mean connection we have:

$$\overline{\nabla}_\gamma \tilde{g}^{\mu\nu} = \frac{\partial \tilde{g}^{\mu\nu}}{\partial x^\gamma} + \tilde{g}^{\alpha\nu}\overline{\Gamma}^\mu_{\alpha\gamma} + \tilde{g}^{\mu\alpha}\overline{\Gamma}^\nu_{\gamma\alpha}.$$

Therefore, for this tensor too, there is an equation of the type (12.4):

$$\overline{\nabla}_\gamma \tilde{g}^{\mu\nu} = \tilde{g}^{\alpha\nu}S^\mu_{\alpha\gamma} + \tilde{g}^{\mu\alpha}S^\nu_{\gamma\alpha}.$$

By decomposing $\tilde{\mathbf{g}}$ into symmetric and antisymmetric parts, this equation is equivalent with the following:

$$\overline{\nabla}_\gamma \tilde{a}^{\mu\gamma} = \tilde{h}^{\mu\alpha}S^\nu_{\gamma\alpha} - \tilde{h}^{\alpha\nu}S^\mu_{\alpha\gamma}$$

which, by contraction over indices γ and ν, gives the result:

$$\overline{\nabla}_\gamma \tilde{a}^{\mu\gamma} = \tilde{h}^{\mu\alpha}S_\alpha; \quad S_\alpha \equiv S^\gamma_{\gamma\alpha}. \tag{12.9}$$

Here S_α a are the components of the *torsion covector*. Now, in the symmetric connection, we have for the covariant derivative of the tensor $\tilde{\mathbf{a}}$ given by

$$\overline{\nabla}_\gamma \tilde{a}^{\mu\nu} = \frac{\partial \tilde{a}^{\mu\nu}}{\partial x^\gamma} + \tilde{a}^{\alpha\nu}\overline{\Gamma}^\mu_{\alpha\gamma} + \overline{\Gamma}^\nu_{\gamma\alpha}\tilde{a}^{\mu\alpha}$$

so that the left hand side of (12.9) is

$$\overline{\nabla}_\gamma \tilde{a}^{\mu\gamma} = \frac{\partial \tilde{a}^{\mu\gamma}}{\partial x^\gamma} + \tilde{a}^{\mu\alpha}\overline{\Gamma}_\alpha \quad \text{with} \quad \overline{\Gamma}_\alpha = \overline{\Gamma}^\gamma_{\gamma\alpha}. \tag{12.10}$$

One can show by routine calculations that this last contraction of the mean connection coefficients amounts to the known form

$$\overline{\Gamma}_\alpha = \frac{\partial}{\partial x^\alpha}\ln\sqrt{|g|}$$

so that the equation (12.10) becomes

$$\overline{\nabla}_\gamma \tilde{a}^{\mu\gamma} = \frac{1}{\sqrt{|g|}}\frac{\partial}{\partial x^\gamma}\left(\tilde{a}^{\mu\gamma}\sqrt{|g|}\right)$$

and with this, from (12.9) we finally have:

$$\tilde{h}^{\mu\alpha}S_\alpha = \frac{1}{\sqrt{|g|}}\frac{\partial}{\partial x^\gamma}\left(\tilde{a}^{\mu\gamma}\sqrt{|g|}\right). \tag{12.11}$$

Therefore, in order that equation (12.1) has unique solution, the condition of nonsingularity of the fundamental tensor: $g \neq 0$ is insufficient. It has to be supplemented with $\tilde{h} \equiv \det(\tilde{\mathbf{h}}) \neq 0$. But this last condition comes to $h \equiv \det(\mathbf{h}) \neq 0$, with the matrix $\mathbf{h}$ defined by equation (12.2), having the entries defined by equation (12.3). Thus, the necessary and sufficient condition for the compatibility in the sense of Einstein between the fundamental tensor and the space connection, is that *both the fundamental tensor and its symmetric part should be nonsingular.* By this, the metric still plays the fundamental part assigned to it by the general relativistic ideas of Einstein, only now this part is complemented with the part played by some other fields, specifically some Yang-Mills type fields. Let us show this in detail.

If $h \neq 0$, the matrix $\mathbf{h}$ is nonsingular, and one can construct, with the help of $\mathbf{h}^{-1}$, a new matrix $\mathbf{g} \cdot \mathbf{h}^{-1} \cdot \mathbf{g}^t$ which is the inverse of $\tilde{\mathbf{h}}$. Indeed, using the associativity of matrix multiplication with respect to matrix addition, as well as the equation (12.7) which defines the inverse of the fundamental tensor, we have by direct calculation:

$$(\mathbf{gh}^{-1}\mathbf{g}^t)\tilde{\mathbf{h}} = \mathbf{gh}^{-1}(\mathbf{h}\tilde{\mathbf{h}} - \mathbf{a}\tilde{\mathbf{h}}) = \mathbf{g}\tilde{\mathbf{g}}^t = 1$$

thus proving our statement. Consequently, if there is an inverse of the matrix **h**, one can find the torsion covector, starting from equation (12.11), by left multiplication with the matrix $\mathbf{g} \cdot \mathbf{h}^{-1} \cdot \mathbf{g}^t$, exactly as in the case of solving the linear algebraic systems, which gives the final result:

$$S_\alpha = (\mathbf{g}\mathbf{h}^{-1}\mathbf{g}^t)_{\alpha\mu} \frac{1}{\sqrt{|g|}} \frac{\partial}{\partial x^\gamma} \left(\tilde{a}^{\mu\gamma} \sqrt{|g|} \right). \qquad (12.12)$$

This is equivalent to the compatibility condition (12.1).

In order to proceed further, we shall provide now a few necessary formulas related to the concept of *affinor*, used by Chernikov himself in order to deduce his formulas. We use here this concept as little as possible, and only because it is very convenient in the calculations involving the torsion: they are almost 'arithmetical'. Generally the matrix stymbolical calculation is to be preferred instead, as being more expedient. In the Riemannian geometry based upon metric **h**, the metric tensor serves also to raise and lower the indices, and therefore to the construction of the mixed tensors of the second orders. These are the affinors: *matrices of transformation of the vectors into vectors of the same variance, either covariance or contravariance.* The most important among affinors of the Chernikov theory, and the only one besides unit affinor in our context, is the antisymetric affinor $\boldsymbol{\phi}$, with entries:

$$\phi_\alpha^\beta = h_{\alpha\mu} a^{\mu\beta} = a_{\alpha\nu} h^{\nu\beta} \qquad (12.13)$$

where **a** is the antisymmetric part of the fundamental tensor **g**. Using this affinor, we can construct a few important tensors, like the ones encountered in equation (12.12), which are essential in the expression of the torsion covector.

The starting point is the definition of the inverse of the fundamental tensor, namely equation (12.7): $\mathbf{g}^{-1} \equiv \tilde{\mathbf{g}}^t$. We need here **g** and $\mathbf{g}^{-1}$, and these matrices can be constructed using the metric matrix **h** and the affinor $\boldsymbol{\phi}$. We have, by definition:

$$\mathbf{g} \equiv \mathbf{h} + \mathbf{a} = (\mathbf{1} + \boldsymbol{\phi})\mathbf{h};$$
$$\mathbf{g}^t \equiv \mathbf{h} - \mathbf{a} = (\mathbf{1} - \boldsymbol{\phi})\mathbf{h}; \qquad (12.14)$$
$$\mathbf{g}^{-1} = \mathbf{h}^{-1}(\mathbf{1} + \boldsymbol{\phi})^{-1}$$

where **1** is the unit affinor, and we used the second of the formulas from equation (12.13), in order to factor the metric tensor out at the right. Thus we have

$$\tilde{\mathbf{h}}^{-1} \equiv \mathbf{g}\mathbf{h}^{-1}\mathbf{g}^t = (\mathbf{1} - \boldsymbol{\phi}^2)\mathbf{h}. \qquad (12.15)$$

The torsion covector (12.12) can now be written in the form

$$S_\alpha = (\mathbf{1} - \boldsymbol{\phi}^2)_\alpha^\mu \frac{1}{\sqrt{|J|}} D_\gamma \left(\tilde{\phi}_\mu^\gamma \sqrt{|J|} \right) \qquad (12.16)$$

214

where D denotes the covariant derivative, this time in the natural connection of the metric $\mathbf{h}$ — the Christoffel symbols of the second kind of the metric — $J \equiv g/h$, and the affinor $\tilde{\boldsymbol{\phi}}$ is defined by $\tilde{\boldsymbol{\phi}} \equiv \mathbf{h}\tilde{\mathbf{a}}$, according to the first equality in (12.13). Based on (12.14) we can further construct the formulas

$$\mathbf{h}\tilde{\mathbf{g}} = (1 - \boldsymbol{\phi})^{-1}; \quad \mathbf{h}\tilde{\mathbf{g}}^t = (1 + \boldsymbol{\phi})^{-1}. \tag{12.17}$$

It thus becomes apparent one of the main advantages of the calculations with affinors, namely that we can use the usual arithmetic rules, provided we preserve the order of factors in the calculation, of course. Obviously, if we have to do with functions of the same affinor, even this rule is unnecessary. For instance, from (12.17) we have right away:

$$\mathbf{h}\tilde{\mathbf{h}} = \frac{1}{2} \left(\frac{1}{1 - \boldsymbol{\phi}} + \frac{1}{1 + \boldsymbol{\phi}} \right) (1 - \boldsymbol{\phi}^2)^{-1}$$

which results from (12.15), by inversion and left multiplication by $\mathbf{h}^{-1}$. By the same token, starting from (12.17) we also have:

$$\tilde{\boldsymbol{\phi}} \equiv \mathbf{h}\tilde{\mathbf{a}} = \boldsymbol{\phi}(1 - \boldsymbol{\phi}^2)^{-1}; \quad \mathbf{a}\tilde{\mathbf{a}} = \boldsymbol{\phi}^2(1 - \boldsymbol{\phi}^2)^{-1}$$

so that as a concrete example of using these relations, the equation (12.16) can also be written as:

$$S_\alpha = \left(\frac{1 - \boldsymbol{\phi}^2}{\sqrt{|J|}} \right)^\mu_\alpha D_\gamma \left[\left(\frac{\boldsymbol{\phi}}{1 - \boldsymbol{\phi}^2} \right)^\gamma_\mu \sqrt{|J|} \right]. \tag{12.18}$$

Now, in the specific four-dimensional case, it is obvious that vanishing of the torsion covector means the vanishing of covariant derivative in either equation (12.12) or equation (12.16), and Chernikov (*loc. cit.*) shows that this further means the Born-Infeld equations of nonlinear electrodynamics.

However, we are interested here in pursuing the idea of continuity, and for this we insist upon another feature of Chernikov's approach: *independence of the theory of the ambient space dimension*. The same algebraical course goes for any space dimension, the details of calculations being decided only by the properties of the affinor $\boldsymbol{\phi}$. Indeed, in order to find the affinor under the covariant derivative in (12.16), which is the affinor designated by symbol $\tilde{\boldsymbol{\phi}}$, Chernikov uses the characteristic equation of the affinor $\boldsymbol{\phi}$, with the final result:

$$\frac{1}{1 - \boldsymbol{\phi}^2} = \frac{Q(\boldsymbol{\phi}^2)}{J} \quad \therefore \quad \tilde{\boldsymbol{\phi}} = \frac{\boldsymbol{\phi}Q(\boldsymbol{\phi}^2)}{J}$$

where Q is a polynomial of degree (n-2)/2 if the dimension 'n' of the ambient space is even, or of the degree (n-3)/2 if the dimension 'n' is odd. This is the only detail of calculation imposed by the dimension of ambient space. The polynomial Q arises as follows: *the Hamilton-Cayley equation* for the affinor $\boldsymbol{\phi}$ assumes one of the forms

$$\mathbf{P}(\boldsymbol{\phi}^2) = \mathbf{0} \quad \text{or} \quad \boldsymbol{\phi}\mathbf{P}(\boldsymbol{\phi}^2) = \mathbf{0} \tag{12.19}$$

as the dimension of the space is even (n = 2k), or odd (n = 2k+1). Therefore P is a polynomial of degree 'k' in its argument. From this it follows that:

$$Q(\boldsymbol{\phi}^2) \equiv \frac{P(1) - P(\boldsymbol{\phi}^2)}{1 - \boldsymbol{\phi}^2} \tag{12.20}$$

for $P(1) - P(\lambda^2)$ is always divisible with $1 - \lambda^2$, by Bézout's classical theorem. Consequently (12.18) can be written in the form:

$$S_\alpha = \left(\frac{\sqrt{|J|}}{Q(\boldsymbol{\phi}^2)} \right)^\mu_\alpha D_\gamma \left[\frac{\boldsymbol{\phi} Q(\boldsymbol{\phi}^2)}{\sqrt{|J|}} \right]^\gamma_\mu \tag{12.21}$$

with $Q(\boldsymbol{\phi}^2)$ calculated as shown in equation (12.20). We shall illustrate right away the case n = 3, for this is *fundamental for a Hertzian theory of matter*.

Chernikov's Theory in the Three-Dimensional Case

From the point of view of the matrix algebra the three-dimensional case of space is detaching among others by a specific property that cannot be found in general for any other dimension. Namely, any antisymmetric 3×3 matrix $\mathbf{a}$ is equivalent to a vector, $\mathbf{v}$ say, by the well-known relation: $a_{ij} = \varepsilon_{ijk} v^k$. Here ε_{ijk} is the totally antisymmetric symbol of Levi-Civita. Thus we can define the square 3×3 matrices necessary to Chernikov's theory:

$$\mathbf{a} \equiv \begin{pmatrix} 0 & v^3 & -v^2 \\ -v^3 & 0 & v^1 \\ v^2 & -v^1 & 0 \end{pmatrix} \; ; \quad \tilde{\mathbf{a}} \equiv \begin{pmatrix} (v^1)^2 & v^1 v^2 & v^1 v^3 \\ v^1 v^2 & (v^2)^2 & v^2 v^3 \\ v^1 v^3 & v^2 v^3 & (v^3)^2 \end{pmatrix} . \tag{12.22}$$

Here $\tilde{\mathbf{a}}$ is the matrix of minors of $\mathbf{a}$, having the entries $a^{ij} = v^i v^j$. The eigenvalue equation of $\mathbf{a}$ in the metric $\mathbf{h}$ can therefore be written in the form:

$$\det(\lambda \mathbf{h} - \mathbf{a}) \equiv h\lambda^3 - \mathrm{tr}(\tilde{\mathbf{h}} \mathbf{a})\lambda^2 + \mathrm{tr}(\mathbf{h}\tilde{\mathbf{a}})\lambda - a = 0$$

where 'h' and 'a' denote the determinants of the two matrices, as usual. Now, because $\mathbf{h}$ is symmetric and $\mathbf{a}$ is antisymmetric, we have a = 0, $\mathrm{tr}(\tilde{\mathbf{h}}\mathbf{a}) = 0$ and

216

$tr(\mathbf{h\tilde{a}}) = h_{ij}v^i v^j \equiv \mathbf{v}^2$, i.e the squared length of the vector $\mathbf{v}$ in the metric $\mathbf{h}$. Therefore the eigenvalue equation reduces to

$$\lambda[\lambda^2 + (\mathbf{v}^2/h)] = 0$$

so that Chernikov's affinor $\boldsymbol{\phi}$ satisfies the Hamilton-Cayley equation:

$$\boldsymbol{\phi}[\boldsymbol{\phi}^2 + (\mathbf{v}^2/h)\mathbf{1}] = \mathbf{0}.$$

This equation defines the polynomial P from the second of the equations (12.19) by

$$P(\boldsymbol{\phi}^2) \equiv \boldsymbol{\phi}^2 + (\mathbf{v}^2/h)\mathbf{1} \quad \therefore \quad \frac{P(1) - P(\boldsymbol{\phi}^2)}{1 - \boldsymbol{\phi}^2} \equiv Q(\boldsymbol{\phi}^2) = 1.$$

In this case, the equation (12.21) becomes:

$$S_k = \sqrt{|h + \mathbf{v}^2|} \cdot D_i \left(\frac{\phi_k^i}{\sqrt{|h + \mathbf{v}^2|}} \right) \qquad (12.23)$$

where '$\mathbf{v}^2$' is, as we said, the magnitude of $\mathbf{v}$, calculated with the metric $\mathbf{h}$.

Now, assume that the metric tensor is brought in actuality by a deformation represented, as known by a symmetric tensor, which we denote by $\mathbf{s}$. Therefore the actual metric tensor will be

$$\overline{\mathbf{h}} = \mathbf{h} + \mathbf{s}.$$

Correspondingly, the mean connection will be of the form:

$$\overline{\Gamma}_{ij}^k = \{_{ij}^k\} + P_{ij}^k \qquad (12.24)$$

where we used the established notation $\{...\}$ for the Christoffel's symbols of the second kind of the metric $\mathbf{h}$, and $\mathbf{P}$ is a tensor part of the connection, symmetric in the lower indices. Therefore the general connection of the actual state will also contain the torsion, so that we shall have:

$$\Gamma_{ij}^k = \overline{\Gamma}_{ij}^k + S_{ij}^k = \{_{ij}^k\} + T_{ij}^k \qquad (12.25)$$

with an obvious definition of the tensor $\mathbf{T}$: it is composed of the tensor $\mathbf{P}$, symmetric in the lower indices, and the torsion tensor $\mathbf{S}$, antisymmetric in the lower indices. We try to solve the compatibility equation in terms of tensor $\mathbf{T}$. Taking (12.25) into the compatibility relation (12.1), results in

$$g_{lk}T_{ij}^l + g_{il}T_{kj}^l = a_{ijk}; \quad a_{ijk} \equiv D_k a_{ij}$$

217

where D denotes, the covariant derivative in the natural connection of the metric **h**, as before. Using here the decomposition (12.2) of the fundamental tensor, results in:

$$T_{ikj} + T_{kij} + \phi_i^l T_{kjl} - \phi_j^l T_{ikl} = a_{ijk}$$

where for the definition of the affinor $\boldsymbol{\phi}$ the metric tensor **h** is used as in equation (12.13). This tensor is also used for lowering and raising the indices of **T**. Now, by transposing the indices 'i' and 'j' the last equation results in

$$T_{jki} + T_{kji} + \phi_j^l T_{kil} - \phi_i^l T_{jkl} + a_{ijk} = 0 \qquad (12.26)$$

from which we have right away:

$$2T_{(ijk)} + a_{(ijk)} = 0 \qquad (12.27)$$

where, for instance, $T_{(ijk)}$ means the result of averaging the tensor over the even permutations of the three indices:

$$3T_{(ijk)} = T_{ijk} + T_{kij} + T_{jki}.$$

From equations (12.26) and (12.27), we have

$$T_{ijk} = a_{ijk} + \phi_j^l T_{kil} - \phi_i^l T_{jkl} - (3/2)a_{(ijk)} \qquad (12.28)$$

which is equivalent to (12.1) for the three-dimensional case.

Decompose now **T** in its symmetric and antisymmetric parts in the first two indices:

$$2P_{ijk} = T_{ijk} + T_{jik}; \quad 2S_{ijk} = T_{ijk} - T_{jik}.$$

This notation is selfexplanatory: raising the last of their indices gives the tensors **P** and **S** defined in equations (12.24) and (12.25) . In terms of these tensors the equation (12.28) itself can be split in two parts:

$$P_{ijk} = \phi_j^l S_{kil} - \phi_i^l S_{jkl} \qquad (12.29)$$

and respectively

$$S_{ijk} = a_{ijk} + \phi_j^l P_{kil} - \phi_i^l P_{jkl} - (3/2)a_{(ijk)} \qquad (12.30)$$

which are still equivalent with the Einstein compatibility condition (12.1). The equation (12.29) shows that the tensor **P** can be calculated once we know the antisymmetric part of the fundamental tensor and the torsion tensor **S**. One essential part of the Chernikov's theory is that the torsion tensor itself can be calculated using only the antisymmetric part of the fundamental tensor. This can be shown as follows: replace $\mathbf{P}$ from equation (12.29) in equation (12.30), and we have as result an equation for the torsion tensor:

$$S_{ijk} + \phi_i^l \phi_k^m S_{ljm} + \phi_j^n \phi_k^m S_{inm}$$
$$= a_{ijk} - (3/2)a_{(ijk)} + \phi_i^m \phi_j^n (S_{nkm} + S_{kmn}).$$

As from (12.29) and (12.30) follows right away that

$$P_{(ijk)} = 0; \quad 2S_{(ijk)} + a_{(ijk)} = 0,$$

the equation in S thus obtained becomes:

$$S_{ijk} + \phi_i^m \phi_j^l S_{mlk} + \phi_i^m \phi_k^l S_{mjl} + \phi_j^n \phi_k^l S_{inl}$$
$$= a_{ijk} - (3/2)[a_{(ijk)} + \phi_i^m \phi_j^n a_{(mnk)}] \tag{12.31}$$

which accounts for our statement, and implicitly for the result of Chernikov's theory taken here as essential: *for the determination of the torsion tensor the fundamental tensor is sufficient.* Therefore, instead of solving the equation (12.1) for the 27 entries of $\mathbf{\Gamma}$, we have to solve in turn the equation (12.31) for the 9 entries of $\mathbf{S}$, then , using $\mathbf{S}$, solve (12.29) for the 18 unknown entries of $\mathbf{P}$, and finally calculate $\mathbf{\Gamma}$ from equation

$$\Gamma_{ij}^k = \{{}_{ij}^k\} + P_{ij}^k + S_{ij}^k$$

resulting from (12.25). Practically we have to 'solve' only the linear system (12.31) for the 9 components of the torsion: the rest of unknowns are simply 'calculated' from these.

However, not just this should be the idea we need to emphasize here, but one relating to the torsion covector $\mathbf{S}$, which results from it: this covector is well defined only by the right hand side of the equation (12.31) . Inasmuch as what we transmitted from Chernikov's theory up to this point, should have suggested the idea of its construction, we just need to reproduce the result we are interested in, by taking it from the original works [(Chernikov, 1976, 1977); see, for instance, (Chernikov, 1976c), p. 13]. Using the metric tensor $\mathbf{h}$ for raising of the last index in equation (12.31) , and the equation (12.21), Chernikov offers an equation for the torsion covector, which we adapt here for the three-dimensional case based on the fact that $Q(\boldsymbol{\phi}^2)$ is the constant polynomial that can be identified with the affinor $\mathbf{1}$:

$$S_k = [1 + \boldsymbol{\phi}^2 / J]_i^j \Psi_{kj}^i;$$
$$\Psi_{kj}^i \equiv \{a_{kjm} - (3/2)[a_{(kjm)} + \phi_k^n \phi_j^p a_{(npm)}]\} h^{mi}. \tag{12.32}$$

If the antisymmetric part of the fundamental tensor is a *vortex*, represented by the curl operation, then $a_{(ijk)} = 0$, so that the tensor $\boldsymbol{\Psi}$ itself can be simply obtained by raising an index with the help of the contravariant metric tensor corresponding to $\mathbf{h}$ by the usual inversion of this matrix:

$$\boldsymbol{\Psi}^i_{kj} = a_{kjm}h^{mi}. \tag{12.33}$$

The point of this whole exercise is in a comparison between equations (12.16) and (12.32) : the structure of torsion is operationally defined by the actual metric tensor and the antisymmetric part of the fundamental tensor. In a word, the fundamental tensor of the three-dimensional space should contain the whole information necessary to transform the space into a three-dimensional phase space. This is, in fact, just the essential idea of the axiomatics of Carlton Frederick, which, we may say, inspired many stochastic approaches of the modern theory of matter in space (Frederick, 1976). Let us stop here for rounding up some conclusions and a briefing on what we are about to discuss further.

Conclusions: Concept of Interpretation and Necessary Further Elaborations

We have reached now a situation where a summary is needed, to set things in order for carrying this work over to a next level, of more technical, in fact very special, mathematical developments for the necessities of Scale Relativity Physics. We just hope that the previous elaboration makes it clear that this theory is a necessary turn of the theoretical physics at large, whereby a few general touches are, nevertheless, indispensable. The work thus far, concerns, even though not exclusively, the 'meta-theoretical' aspect of SRT mentioned by Laurent Nottale in the concluding words of his own work, reproduced by us in the Introduction (Nottale, 2011): SRT is, in fact, a necessary *theory of interpretation*. With the advent of wave mechanics, the interpretation has been established as a necessity within the core of physics itself. Our analysis shows that interpretation is necessary even in the classical physics: assuming, for instance, that it is possible to eliminate the wave mechanics from our knowledge, the interpretation still cannot be eliminated. Its permanence within human knowledge is formally assessed by the connection with the mathematical theory of ensembles, which in turn involves the general idea of density, at least as long as physics is concerned.

Thus, we have chosen in this work a few fundamental historical points of crucial adjustment of the human knowledge — 'moments of knowledge', as we would like to call them — having a precise criterion of selection in mind: those moments destined *to turn the idea of interpretation into a well-defined concept*. For, with the advent of the wave mechanics, the idea of interpretation became critical only by the disclosure of its object: translation of the theoretical terms, mathematically expressed by the concept of wave function, into 'experimental' terms, mathematically expressed by the concept of particle (Darwin, 1927). The construction of the concept of interpretation itself involves that 'refoundation on mathematical principles' mentioned by Nottale in the excerpt from our Introduction here. On this key note, the fundamental feature of the concept of interpretation turns out to be the idea of freedom of particles: *the wave function is referring to an ensemble of free particles*. Once realizing this, a host of classical mechanical concepts fall into their right places, and the historical development of the physics since the birth of the two new mechanics — undulatory and quantal — helps in making this idea more precise. Hence our choice for a few significant 'moments of knowledge'.

The freedom here is not classical, involving, as it were, only the principles of dynamics, but rather wave mechanical: the wave function describes an ensemble under no potential. This was suggested by Erwin Madelung, who constructed a first ensemble for the concept of interpretation of wave mechanics (Madelung, 1927). On this note, we present the potential itself as purely quantal in nature, no matter of the scale of space and time. This conclusion is derived directly from the idea of interpretation which, while based on the *nonstationary Schrödinger equation with no potential*, as a fundamental mathematical tool of the process of interpretation, also imposes a special position of the potential: it is correlated with the behavior of the amplitude of the wave function described by a *stationary Schrödinger equation*. In this presentation the stationary Schrödinger equation is not essentially an eigenvalue equation, but only an equation to be satisfied by the amplitude of the complex wave function. Thus, the classical aspect of freedom results in a special choice of the amplitude of the wave function, which turns out to be compatible with the Louis de Broglie's ideas deriving from the association of a frequency to a massive particle using the concept of energy. This fact was shown by Lachlan Mackinnon, who pointed out the way in which a de Broglie wave is related to the *measurement of phase*, considered as a fundamental proces in an enclosure (Mackinnon, 1978). A lighter note for the concept of interpretation, but perhaps by far more popular in the present state of our minds, is then revealed as a task of the wave mechanics: this physical science concerns the reconciliation between the de Broglie and Schrödinger moments of our knowledge. Perhaps now it is the right occasion for a brief summary of those two moments, via the very accomplishments of the two great scholars, as described in this work.

We uphold the idea that the root of the concept of interpretation may have been the classical theory of color vision, which, by the beginning of the last century has reached a Riemannian geometrical status, especially through the works of Hermann von Helmholtz. For, it was Schrödinger who, in his essential work on the theory of color vision (Schrödinger, 1920), took notice of the critical need of the theory of color vision. This was the connection with the Cantorian theory of continua, necessary for the arithmetization of colors, and this is basically a theory of ensembles in the modern understanding. With Schrödinger an essential idea achieves a precise contour in this moment: while Riemannian extended manifolds of color may have an infinite dimension, even a continuous dimension at that, their physical 'qualities' are finite dimensional manifolds of maximal dimension three. We found it significant for this moment of our knowledge, that the year 1920 seems to have been a turning point for the personality of Schrödinger: for the rest of his life he dedicated himself to theoretical physics and natural philosophy. So, in our opinion, the celebrated works from 1926, in which Schrödinger introduced the concept of wave function, are a direct consequence of this turn of his life, prompted by the idea of color vision. For, the wave function as introduced by Schrödinger represents nothing else but the possibility of conceiving simultaneously an ensemble of material points having one of the powers of continua.

If we are to continue on the same note, we can say that Louis de Broglie pressed on an apparently more objective concept of physical science related to the issue at hand: the energy. In mechanics, the energy of a complex system is a vague concept, to say the least (Poincaré, 1897). With statistical mechanics, it became a matter of statistics, for a few cases where we can claim a definition of some kinds of measurable energy, at any rate. In one such case, namely that of the 'wave phenomenon called classical material point', in the terminology of de Broglie, a frequency can be associated to a material point (de Broglie, 1923), using the precepts of special relativity. But, in this instance, the frequency necessarily plays a twofold part: first as an expression of the energy indeed, then as an expression of time in a process of measurement of the phase of wave phenomenon. According to relativity this leads to contradiction, which in turn led de Broglie to assume that 'the wave phenomenon called material point' is actually a *group of waves*, all 'in phase'. This is the origin of the whole of de Broglie's work, and certainly that of the subsequent construction of Lachlan Mackinnon illustrated in this work.

Thus, the 'de Broglie moment' was characterized here by two essential accomplishments related to the concept of interpretation: the very first example of a Madelung ensemble of free material points, necessary for the interpretation of a classical concept – that of physical light ray (de Broglie, 1927) – and a classical manner of application of field on matter and of matter on field (de Broglie, 1935). While both these concepts are of a classical origin, they are nevertheless universal in the general economy of the concept of interpretation. They both revealed the tremendous role played by the physical theory of surfaces in the process of interpretation, a role that ended up in the future concept of holography. In particular, only this way we are able to state that the density of an ensemble is related to the square of the amplitude of the wave function interpreted by this ensemble. We might rightfully say that this is a 'holographic' property, and that Louis de Broglie added in fact to the classical concept of ray a necessary differentia left unfinished by the physical optics of Augustin Fresnel. This is the connection of the local description of physical ray with the global description of light by a wave surface, accomplished by Fresnel only as a transition between infrafinite and finite orders, in mathematical terminology. Louis de Broglie shows in fact that this should be a universal connection for the idea of scale transition, making out of special relativity a scale relativity in the sense of Laurent Nottale.

Thus, in order to be able to merge these two moments of our knowledge — de Broglie and Schrödinger — we followed the idea of which Laurent Nottale himself took notice, namely of "rendering the whole physics relativistic". Only, we chose for this job the quintessential model of a physical system, chosen, as a matter of fact, by Schrödinger himself for illustrating the virtues of wave function in a problem with eigenvalues: the classical hydrogen atom. This model contains naturally the idea of limitation of velocities, which is crucial for the kind of matter interpreted by light: the ether. Incidentally, by this the ether becomes itself the quintessential matter, with photons as ultimate material particles. Then, we just added space extension to the physical components of the classical planetary atom, a notion for which we needed a 'twist' on the classical natural philosophy, due to Heinrich Hertz. According to Hertz, the classical material point is endowed with a feature associated to the concept of time in special relativity, and thus becomes a *Hertz material particle*: a means of associating with one another different positions in space at different time moments (Hertz, 2003). By this the freedom conferred to material particles in a Schrödinger approach can be classically assessed as the freedom of particles in matter, whose expression is similar to the contemporary idea of *asymptotic freedom*. In our general connotation, the freedom of material particles in matter is a classical freedom of the material point under forces in equilibrium, but in a limited space, not in the universe; or, we might say, the freedom of a Hertz material particle in the universe momentarily at its disposal. This property is what makes the description of matter a ... matter of scale, and asks for a necessary development of theoretical physics into a Scale Relativity Physics.

Thus, coming back to the idea of scale transition, both in time and space, another moment of knowledge needs to be considered, and we designated it as the 'Berry moment', forasmuch as it is related to the name of Sir Michael Berry. Three essential achievements were then associated by us with this moment of knowledge, each one of them regarding exclusively the transition between scales of space and time. The first one of these achievements is referring to idea of evolution of the phase of wave function, whereby the space dependence of phase is implicit in the phase of phenomena described adiabatically (Berry, 1984). We have insisted here on the fact that this achievement of Berry allows actually a geometrical description of the adiabatical parameter space. This fact is only implicit in the original works of quantum mechanics [see (Heisenberg, 1925)], and cannot be made explicit but only by interpretation, thus involving the idea of wave function: hence the necessity of a geometry of parameters. The second achievement listed by us under 'Berry moment' of knowledge is the liberation of the geometry of parameters' space from the restriction of adiabaticity (Berry & Klein, 1984). This is accomplished by the idea of a time dependent space scale transition which leaves the Newtonian forces unchanged. For once, the procedure has the general cosmological connotation of the adiabatic procedure of transition, which leaves the Planck spectrum invariant. Thus, we are guided to the conclusion that the Newtonian type of forces are cosmological in nature. We only point out an essential difference here: while the Planck spectrum represents, *in a mathematical time order*, a direct transition between infrafinite and transfinite, the Newtonian forces represent a transition between infrafinite and finite. In general, therefore, the independence with respect to the rate of processes involved in interpretation is to be connected to the invariance to scale changes. Finally, the third achievement listed by us under the 'Berry moment' of knowledge is what we termed as the 'Airy moment' of the wave function (Berry & Balazs, 1979). Thereby, the wave function can be expressed by an integral equation having a solution of the nonstationary Schrödinger equation as kernel acting on Airy functions in space. We correlated this situation with the third order nonstationary equation in space variable, which, in the Nottale's approach of the fractal fluid structure would mean dispersion of waves. Classically, such a linear equation is equivalent to an integral equation, having nevertheless the Airy function as a kernel (Widder, 1979). So, the lesson we extracted from this 'Airy moment' can be expressed in terms of Louis de Broglie's idea of 'application of field on a particle' and of 'a particle on the field' via some physical quantities.

However, this last contribution to the 'Berry moment' has by far much more significance for human knowledge in general, which strikes the eyes when noticing an ubiquitous implicit dependence of wave function and potential on space and time. Specifically, they depend on space and time not directly, but via some algebraic expressions representing physical situations. Originally this behavior has been relegated by Berry and Balazs to the onset property of the Airy function, of describing the caustics of light rays. This way the behavior can be transferred to a phase space property, thus describing an ensemble of particles. While maintaining this original idea, in the present work we pressed on another interpretation (Greenberger, 1980), by the equivalence principle: namely, here we have to do with an ensemble of free Hertz particles in matter, describing an Einstein elevator in free fall, to wit, a space extended particle in a gravitational field. In this interpretation, a special statistics of the Planck type is used for the time moments, which should thus be submitted to the same type of quantization as the spectral energy of light. For once, this testifies for the physical identity between a Wien-Lummer enclosure and an Einstein elevator, thus pointing out towards an identity between the two implements of our practical knowledge of the world. However, the implication of the 'Airy moment' of the wave function are by far more intricate and affluent in consequences.

While these consequences can be a subject matter of substantial development of the present ideas, we only need to show here that our approach of the interpretation concept − whereby the wave function is interpreted by ensembles of Hertz free particles, and the potential of forces describes implicitly these forces − is a natural situation according to the very idea of quantization of the wave mechanics. In hindsight, this interpretation warrants the Nottale's approach of SRT by geodesic motions description of a fractal fluid.

The idea of quantization of the wave mechanics starts from postulating the Schrödinger equation in the form

$$i\hbar\frac{\partial\Psi}{\partial t} = \mathbf{H}\Psi$$

where $\mathbf{H}$ is the Hamiltonian operator. Admitting autonomy for the wave function, $\mathbf{M}\Psi$ can be very well another valid wave function, satisfying the Schrödinger equation

$$i\hbar\frac{\partial(\mathbf{M}\Psi)}{\partial t} = \mathbf{H}(\mathbf{M}\Psi)$$

for the same Hamiltonian. Here $\mathbf{M}$ is an operator acting on the wave function in order to construct another wave function. It is then easy to see that this operator should be constrained to satisfy the nonstationary Heisenberg-Dirac type equation:

$$i\hbar\frac{\partial\mathbf{M}}{\partial t} = [\mathbf{H}, \mathbf{M}].$$

Such an operator can be very well that involved in an eigenvalue equation, whereby the wave function thus defined appears as the original wave function multiplied by a constant:

$$\mathbf{H}_0\Psi = E\Psi.$$

Thus, if the wave function of the eigenvalue problem is to be interpreted by an ensemble, $\mathbf{H}_0$ has to satisfy the Heisenberg-Dirac equation of motion

$$i\hbar\frac{\partial\mathbf{H}_0}{\partial t} = [\mathbf{H}, \mathbf{H}_0]$$

and if the operator is of the particular form

$$\mathbf{H}_0 = -\frac{\hbar^2}{2m}\Delta + V$$

this evolution equation turns out to be an evolution equation for the potential only. According to Lax's theorem, which is the basis of the theory of solitons (Lax, 1968), E has to be then an integral of the evolution described by $\mathbf{H}$, depending only implicitly on time, while $\mathbf{H}\Psi$ itself must be a solution of the eigenvalue equation for $\mathbf{H}_0$. This can be seen rather directly, by differentiating the eigenvalue equation for $\mathbf{H}_0$, while assuming an explicit dependence of time for the wave function. In the conditions just stated, the result is

$$\left\{ \frac{\partial \mathbf{H}_0}{\partial t} - \frac{1}{i\hbar}[\mathbf{H}, \mathbf{H}_0] - \frac{\partial E}{\partial t} \right\} \Psi = 0$$

showing that the evolution of the wave function is compatible with the eigenvalue problem only if E does not depend on time. The question is, what kind of dependence? Is this time the time of an *evolution* or the time of *motion*?

The discrimination between these two aspects of time makes all the difference within the concept of interpretation. For, if the time is a continuity parameter, we can talk about an ensemble of Hertz material particles corresponding to a moment of time indeed. It is such an ensemble that can be described by a Planck statistics. On the other hand, if the time is a parameter of motion, it is referring to a single Hertz material particle. In the first case the time derivative is a partial derivative, while in the second case we have a total derivative expressed as a Lie transport along the motion or, in general, along a vector related to this motion. This is the case to which the Nottale's complex fluid is referring. However, within the concept of interpretation, the nonstationary Schrödinger equation involves the first aspect of time, which in turn involves the partial derivative. Therefore the energy as an eigenvalue may not be constant, but can depend implicitly of time, as the 'Airy moment' plainly shows indeed.

This fact served us as a lesson about conceiving the freedom of the Hertz particles in matter, by a generalization of the concept of adiabaticity along the ideas resuscitated by the 'Berry moment': the motion of the material particles in matter is much faster than the evolution of ensembles of particles perceived by our senses as motion at a certain space scale. In other words, the evolution can always be described as an adiabatic motion, but the very notion of 'adiabatic' is a matter of scale. It was Louis de Broglie who implicitly suggested such a condition, as the instrumental condition for expressing the density of matter by the square of the amplitude of the wave function. In a word, the conclusion of de Broglie's analysis amounts to the general statement that the motion of particle to which a frequency can be associated is adiabatic with respect to the speed of possible representative waves. This clarifies the position of the potential with respect to the wave function: the amplitude of this function depends on space and time only implicitly, through the potential. The plane waves and the Airy wave functions are just particular cases of this general situation, which can be characterized along the lines of comparison made by Synge, between the solution of the nonstationary Schrödinger equation and the de Broglie wave packets (Synge, 1972). The bottom line here is that a proper characterization of the concept of interpretation involves in turn the concept of density of matter in its utmost generality: Newtonian, as a characteristic of continuity of the matter in space, together with Einsteinian as a cardinality of the ensemble of Hertz particles interpreting the matter.

At this juncture we adopted the geometrical views of David Delphenich, according to which the *density of matter is related to the torsion* of the Riemannian space representing it (Delphenich, 2013). This turns out to be what we think as the right way to describe the geometry of ensembles, when combined with an old Cartanian view on the distant parallelism serving the purposes of the general theory of relativity (Cartan, 1931). To wit, it allows us to describe the matter as filling an Euclidean space, by physical methods involving the propagation of signals, whereby the torsion is related to the fundamental fields.

So much for the concept of interpretation contained in this work. Of course the work itself includes more than this, mostly along the idea of procedures of describing the continuity related to the density, especially an original update on the *transport theory*. However, the specific technicalities for developing SRT, still remain to be expounded, and will be listed here in no particular order, by the way of conclusion, for the benefit of a proper perspective. They should be as follows.

A characterisation of the Hertz material particle within matter, the way Schrödinger describes the color: the material points are interpreted as ensembles having the powers of continua, yet the Hertz material particles inside material points are defined through three qualities: the mass and the two charges, electric and magnetic, in a static definition of the material point, based on the simplest second principle of dynamics (Wigner, 1954). This principle sounds: *under no forces a particle is at rest.* The approach results in a non-Euclidean theory of mass and charges, allowing a specific correspondence between positions in space and Hertz material particles, left unspecified in the definition of Heinrich Hertz for a material particle, but contained implicitly in the Newtonian formulation of the dynamics. It should be our task to describe this physics in detail, with a 'refoundation' based on 'mathematical principles', using the Nottale's own expression.

The rate independence of the general adiabatic processes as defined here, which is an essential aspect arising with the 'Berry moment' of knowledge, reveals the fundamental role played by the harmonic oscillators in the transitions of space scale. This aspect needs to be deepened and further elaborated, along the idea of statistics related to space variables. The transition between space scales selects certain space variables. It turns out that this transition generates a statistics of the same nature as the time one: *a Planck statistics*. The analysis should allow a proper characterization of the geometry, based on a natural concept of physical reference frame.Incidentally, some other concepts of reference frame follow naturally along the developments.

Of course, we need to be concerned with the universalization of relativity, as requested by Laurent Nottale. As it turns out, this occurs in both aspects of the relativity: special and general. It should allow us to reenact an old attempt of constructing the general relativity starting directly from special relativity, within the analysis directly related to the planetary model. All these technicalities, once rounded up, allow, in our opinion, a corresponding necessary round up of the Scale Relativity Physics itself. The first thing to show in this respect is a connection between propagation and Schrödinger equation, as the right manner of relating the two aspects of time involved in relativity. This fact will resuscitate an old observation of Harry Bateman, which leads us to a theory of measurement, adopted and adapted from the quantum theory of spin. The theory of surface tension then ensues quite naturally, as the generalization of the idea of quantal spin, in the process of interpretation, in a general 'de Broglie moment' of the wave mechanics. The Scale Relativity in Nottale's approach turns out to be a Lagrangian theory, whereby the Lagrangian has a statistical meaning related to the statistics of time sequences (Dirac, 1933).

References

Agaoka, Y. (1989): *On a Generalization of Cartan's Lemma*, Journal of Algebra, Volume **127**(2), pp. 470 - 507.

Agop, M., Păun, V., Harabagiu, A. (2008): *El Naschie's $\varepsilon^{(\infty)}$ Theory and Effects of Nanoparticle Clustering on the Heat Transport in Nanofluids*, Chaos, Solitons and Fractals, Volume **37**, pp. 1269 - 1278.

Aharonov, Y., Bohm, D. (1959): *Significance of Electromagnetic Potentials n the Quantum Theory*, Physical Review, Volume **115**, pp. 485 - 491.

Airy, G. B. (1838): *On the Intensity of Light in the Neighbourhood of a Caustic*, Transactions of the Cambridge Philosophical Society, Volume **VI**, pp. 379 - 402.

Airy, G. B. (1848): *On the Intensity of Light in the Neighbourhood of a Caustic*, Transactions of the Cambridge Philosophical Society, Volume **VIII**, pp. 595 - 599.

Alfaro, V. de, Fubini, S., Furlan, G. (1976): *Conformal Invariance in Quantum Mechanics*, Il Nuovo Cimento A, Volume **34**, pp. 569 - 611.

Appell, P. (1889): *De l'Homographie en Mécanique*, American Journal of Mathematics, Volume **12**(1), pp. 103 - 114.

Appell, P. (1891): *Sur les Lois de Forces Centrales Faisant Decrire a Leur Point d'Application' une Conique Quelles Que Soient les Conditions Initiales*, American Journal of Mathematics, Volume **13**(2), pp. 153 - 158.

Arnold, V. (1976): *Les Méthodes Mathématiques de la Mécanique Classique*, Editions MIR, Moscou.

Bacry, H. (1988): *The Position Operator Revisited*, Annales de l'Institut Henri Poincaré, Physique Théorique, Volume **49**(2), pp. 245 - 255.

Bacry, H. (1993): *Which Deformations of the Poincar Group?*, Journal of Physics A: Mathematical and General, Volume **26**(20), pp. 5413 - 5425.

Baldwin, P. R., Townsend, G. M. (1995): *Complex Trkalian Fields and Solutions of Euler's Equations for the Ideal Fluid*, Physical Review E, Volume **51**(3), pp. 2059 - 2068.

Barbilian, D. (1937): *Die von Einer Quantik Induzierte Riemannsche Metrik*, Comptes Rendus de lAcadémie Roumaine des Sciences, Volumul **2**, p.

198.

Barbilian, D. (1938): *Riemannsche Raum Cubischer Binärformen*, Comptes Rendus de l'Académie Roumaine des Sciences, Volumul **2**, p. 345.

Barbilian, D. (1971): *Elementary Algebra*, in *Didactic Works of Dan Barbilian*, Volume **II**, Editura Tehnică, Bucureşti (in Romanian).

Baskoutas, S., Jannusis, A. (1992): *Quantum Tunneling Effect for the Inverted Caldirola-Kanai Hamiltonian*, Journal of Physics A: Mathematical and General, Volume **25**, pp. L1299 - L1304.

Bateman, H. (1918): *Differential Equations*, Longmans, Green & Co., London.

Belbruno, I. M. (1977): *Two-Body Motion under the Inverse Square Central Force and Equivalent Geodesic Flows*, Celestial Mechanics, Volume **15**(4), pp. 467 - 476.

Bellman, R. (1997): *Introduction to Matrix Analysis*, Society for Industrial and Applied Mathematics, Philadelphia PA.

Berry, M. V. (1984): *Quantal Phase Factors Accompanying Adiabatic Changes*, Proceedings of the Royal Society of London, Series A, Volume **392**, pp. 45 57.

Berry, M. V. (1985): *Classical Adiabatic Angles and Quantal Adiabatic Phase*, Journal of Physics A: Mathematical and General, Volume **18**(1), pp. 15 27.

Berry, M. V. (1989): *The Quantum Phase, Five Years After*, Original introductory contribution to (Shapere & Wilczek, 1989).

Berry, M. V., Balazs, N. L. (1979): *Nonspreading Wave Packets*, American Journal of Physics, Volume **47**, pp. 264 - 267.

Berry, M. V., Klein, G. (1984): *Newtonian Trajectories and Quantum Waves in Expanding Force Fields*, Journal of Physics A: Mathematical and General, Volume **17**(8), pp. 1805 - 1815.

Betounes, D. E. (1983): *The Kinematical Aspect of the Fundamental Theorem of Calculus*, American Journal of Physics, Volume **51**, pp. 554 - 560.

Bianchi, L. (2001): *On Three-Dimensional Spaces Which Admit a Continuous Group of Motions*, General Relativity and Gravitation, Volume **33**, pp. 2171 - 2253 (This is a translation of the original Italian from year 1897, by R. T. Jantzen under Golden Oldies initiative of the periodical).

Bloore, F. J. (1977): *The Shape of Pebbles*, Mathematical Geology, Volume **9**(2), 113 - 122.

Bohm, D. (1952): *A Suggested Interpretation of Quantum Theory in Terms of "Hidden" Variables I*, Physical Review, Volume **85**(2), pp. 166 - 179; *A Suggested Interpretation of Quantum Theory in Terms of "Hidden" Variables II*, Ibidem, pp. 180 - 193.

Bohr, N. (1913): *On the Constitution of Atoms and Molecules*, The Philosophical Magazine, Volume **26**, pp. 1-24.

Boltyanskii, V.G. (1974): *Anisotropic Relativism* (in Russian), Differential Equations, Volume **10**(11), pp. 2101 - 2110.

Boltyanskii, V.G. (1979): *The Anisotropic theory of Relativity and Optimization* (in Russian), Differential Equations, Volume **15**(10), pp. 923 - 932.

Born, M. (1934): *On the Quantum Theory of the Electromagnetic Field*, Proceedings of the Royal Society of London, Series A, Volume **143**, pp. 410 - 437.

Born, M., Infeld, L. (1934): *Foundations of the New Field Theory*, Proceedings of the Royal Society of London, Series A, Volume **144**, pp. 425 - 451.

Born, M., Oppenheimer, J. R. (1927): *Zur Quantentheorie der Molekeln*, Annalen der Physik, Volume **84**, pp. 457 - 484; *On the Quantum Theory of Molecules*, the English version of the German original, by S. M. Blinder.

Bowman, F. (1958): *Introduction to Bessel Functions*, Dover Publications, New York.

Box, G. E. P., Cox, D. R. (1964): *An Analysis of Transformations*, Journal of the Royal Statistical Society, Series B, Volume **26**(2), pp. 211 - 252.

Boyer, T. H. (1969): *Derivation of the Blackbody Radiation Spectrum without Quantum Assumptions*, Physical Review, Volume **182**(4), pp. 1374 - 1383.

Boyer, T. H. (1975): *Random Electrodynamics: the Theory of Classical Electrodynamics with Classical Electromagnetic Zero-Point Radiation*, Physical Review D, Volume **11**, pp. 790 - 808.

de Broglie, L. (1923): *Ondes et Quanta*, Comptes Rendus de l'Académie des Sciences, Paris, Tome **177**, pp. 507 - 510.

de Broglie, L. (1926): (a) *Sur le Parallélisme entre la Dynamique du Point Matériel et l'Optique Géométrique*, Le Journal de Physique et le Radium, Srie VI, Tome **7**(1), pp. 1 6; (b) *Sur la Possibilité de Relier les Phénomènes d'Interférence et de Diffraction à la Théorie des Quanta de Lumière*, Comptes Rendus de l'Académie des Sciences de Paris, Tome **183**, pp. 447448; (c) *Interference and Corpuscular Light*, Nature, Volume **118**, pp. 441 - 442; (d) *Les Principes de la Nouvelle Mécanique Ondulatoire*, Le Journal de Physique et le Radium, Tome **7**(10), pp. 321 337; (e) *Ondes et Mouvements*, Gauthier-Villars, Paris.

de Broglie, L. (1935): *Une Remarque sur l'Interaction entre la Matière et le Champ Électromagnétique*, Comptes Rendus de l'Académie des Sciences de Paris, Tome **200**, pp. 361 - 363.

de Broglie, L. (1966): *Thérmodynamique de la Particule Isolée ou Thérmodynamique Cachée de la Particule*, Gauthièr-Villars, Paris.

Brown, L. M. (Editor) (2005): *Feynman's Thesis - a New Approach to Quantum Mechanics*, World Scientific, Singapore.

Burbea, J. (1986): *Informative Geometry of Probability Spaces*, Expositiones Mathematicae, Volume **4**, pp. 347 - 378.

Burdet, C. A. (1968): *Theory of Kepler Motion: the General Perturbed Two Body Problem*, ZAMP, Volume **19**, pp. 345 - 368.

Burdet, C. A. (1969): *Le Mouvement Keplerien et les Oscillateurs Harmoniques*, Journal für die Reine und Angewandte Mathematik, Volume **238**, pp. 71 - 84.

Caldirola, P. (1941): *Forze Non Conservative nella Meccanica Quantistica*, Il Nuovo Cimento, Volume **18**(9), pp. 393 - 400.

Caldirola, P. (1983): *Quantum Theory of Nonconservative Systems*, Il Nuovo Cimento B, Volume **77**(2), pp. 241 - 262.

Calmet, J., Calmet, X. (2004): *Metric on a Statistical Space-Time*, WSEAS Transactions on Circuit and Systems, Volume **3**(10), pp. 2267-2271; arXiv:math-ph/0403043v1.

Carinena, J. F., Clemente-Gallardo, J., & Marmo, G. (2007): *Reduction Procedures in Classical and Quantum Mechanics*, International Journal of Geometrical Methods in Modern Physics, Volume **4**, pp. 1363 - 1403; arXiv:math-ph/0709.2366.

Cartan, E. (1930): *Notice Historique sur la Notion de Parallélisme Absolu*, Mathematische Annalen, Tome 102, pp. 698 - 706; English translation in (Delphenich, 2011), pp. 130 137.

Cartan, E. (1931): *Le Parallélisme Absolu et la Théorie Unitaire du Champ*, Revue de Métaphysique Et de Morale, Tome **38**(1), pp. 13 - 28; ; English translation in (Delphenich, 2011), pp. 202 - 211.

Cartan, E. (1951): *La Théorie des Groupes Finis et Continus et la Géométrie Différentielle Traitées par la Méthode du Repère Mobile*, Gauthier-Villars, Paris.

Cartan, E. (2001): *Riemannian Geometry in an Orthogonal Frame*, World Scientific Publishing, Singapore.

Casalis, M. (1996): *The 2d+4 Simple Quadratic Natural Exponential Families on R^d* , The Annals of Statistics, Volume **24**, pp. 1828 - 1854.

Casian-Botez, I., Agop, M., Nica, P., Păun, V., Munceleanu, N. M. (2010): *Conductive and Convective Types Behaviors at Nano-Time Scales*, Chaos, Solitons and Fractals, Volume **7**(11), pp. 1 - 10.

Cayley, A. (1859): *A Sixth Memoir Upon Quantics*, Philosophical Transactions of the Royal Society of London, Volume **149**, pp. 61 - 90. Reproduced in *The Collected Mathematical Works*, Volume II, pp. 561 - 592, Cambridge University Press.

Červený, V. (2005): *Seismic Ray Theory*, Cambridge University Press, Cambridge, UK.

Chandrasekhar, S. (1995):*Newton's Principia for the Common Reader*, Oxford University Press, New York.

Chernikov, N. A. (1976): (a) *The Torsion Covector in the Unitary Field Theory*, (in Russian) Preprint **P2-9651**, Dubna; (b) *The Born-Infeld Equations in Einstein's Unitary Field Theory*, (in Russian) Preprint **P2-9681**, Dubna; (c) *The Born-Infeld Equations in Einsteins Unitary Field Theory*, [in Russian; an extension of the preceding work (b)] Preprint **P2-9714**, Dubna.

Chernikov, N. A. (1977): *Affine Connection Space with Asymmetric Tensor Field*, (in Russian) Preprint **P2-11021**, Dubna.

Clifford, W. K. (1882): *Mathematical Papers*, MacMillan & Comp., London.

Coddington, H. (1829): *A System of Optics*, Volume I, J. Smith, Cambridge, UK.

van Dantzig, D. (1934): *The Fundamental Equations of Electromagnetism, Independent of Metrical Geometry*, Mathematical Proceedings of the Cambridge Philosophical Society, Volume **30**(4), pp. 421 - 427.

Darboux, G. (1877): *Recherche de la Loi que Doit Suivre une Force Centrale pour que la Trajectoire qu'elle Détermine Soit Toujours une Conique*, Comptes Rendus de l'Académie des Sciences Paris, Tome **84**, pp. 760 - 762; 936 - 938.

Darboux, G. (1884): *Sur les Lois de Kepler*, Note XII en Cours de Mécanique par Théodore Despeyrous, Hermann, Paris, Tome I, pp. 432 - 440.

Darwin, C. G. (1927): *Free Motion in the Wave Mechanics*, Proceedings of the Royal Society of London Series A, Volume **117**, pp. 258 - 293.

Delphenich, D. H. (2002): *A Geometric Origin of the Madelung Potential*, **arXiv**:0211065 [gr-qc].

Delphenich, D. H. (Ed) (2011): *Selected Papers on Teleparallelism*, www.neo-classical-physics.info.

Delphenich, D. H. (2013): (a) *A Strain Tensor that Couples to the Madelung Stress Tensor*, **arXiv**: 1303.3582v1 [quant-ph]; (b) *The Use of Teleparallelism Connection in Continuum Mechanics*, **arXiv**: 1305.3477v1 [gr-qc].

Dicke, R. H., Peebles, P. J. E., Roll, P. G., Wilkinson, D. T. (1965): *Cosmic Black-Body Radiation*, The Astrophysical Journal, Volume **142**, pp. 414 - 419.

Dirac, P. A. M. (1931): *Quantised Singularities in the Electromagnetic Field*, Proceedings of the Royal Society of London, Series A, Volume **133**, pp. 60 - 72.

Dirac, P. A. M. (1933): *The Lagrangian in Quantum Mechanics*, Physikalische Zeitschrift der Sowietunion, Volume **3**(1), pp. 64 - 72; reproduced in (Brown, 2005).

Dirac, P. A. M. (1948): *The Theory of Magnetic Poles*, Physical Review, Volume **74**(6), pp. 807-830.

Dirac, P. A. M. (1962): *An Extensible Model for the Electron*, Proceedings of the Royal Society of London, Series A, Volume **268**, pp. 57 - 67.

Dubois-Violette, M., Kerner, R., Madore, J. (1990): *Noncommutative Differential Geometry of Matrix Algebras*, Journal of Mathematical Physics, Volume **31**(2), pp. 316 - 322.

Dumachev, V. N. (2009): *Generalized Nambu Dynamics and Vectorial Hamiltonians*, **arXiv**: mathDG/0904.4326v1.

Dumachev, V. N. (2010): *Lax Pair for 2D System of Linear Differential Equations*, International Journal of Pure and Applied Mathematics, Volume **63**, pp. 207 - 212.

Dumachev, V. N. (2011): *Phase Flows and Vector Hamiltonians*, Russian Mathematics (Iz. VUZ), Volume **55**, pp. 1 - 7.

Duval, C., Guieu, L. (1998): *The Virasoro Group and the Fourth Geometry of Poincaré*, **arXiv**: mathDG/9806135v1.

Eells, J., Sampson, J. H. (1964): *Harmonic Mappings of Riemannian Manifolds*, American Journal of Mathematics, Volume **86**, pp. 109 - 160.

Ehrenfest, P. S. (1927): *Bemerkung über die angenäherte Gültigkeit der klassischen Mechanik innerhalb der Quantenmechanik*, Zeitschrift für Physik, Volume **45**, pp. 455 - 457; *Remark on the Approximate Validity of Classical Mechanics within Quantum Mechanics*, English translation by D. H. Delphenich; www.neo-classical-physics.info.

Einstein, A. (1905): *On the Electrodynamics of Moving Bodies*, in *The Principle of Relativity, a Collection of Original Memoirs on the Special and General Theory of Relativity*, Dover Publications, 1923, 1952, 2003. English translation of the original from Annalen der Physik, Volume **17** (1905).

Einstein, A. (1909): *Planck's Theory of Radiation and the Theory of Specific Heat*, The Collected Papers of Albert Einstein, Volume 2, Princeton University Press; English translation of the original from Deutsche Physikalische Gesellschaft, Verhandlungen **7**, pp. 482 - 500; also published in Physikalische Zeitschrift, Volume **10**, pp. 817 - 825.

Einstein, A. (1930): *Auf die Riemann-Metrik und den Fern-Parallelismus gegründete einheitliche Feldtheorie*, Mathematische Annalen, Tome **102**, pp. 685 - 697; English translation in (Delphenich, 2011), pp. 117 - 129.

Einstein, A. (1949): *Remarks Concerning the Essays Brought Together in this Co-Operative Volume*, in *Albert Einstein, Philosopher-Scientist*, P. A. Schilpp editor; Volume VII, the Library of Living Philosophers, pp. 665 - 688, MJF Books, NY.

Einstein, A. (1965): *Concerning an Heuristic Point of View toward the Emission and Transformation of Light*, American Journal of Physics, Volume **33**(4), pp. 367 - 374; English translation of the original from Annalen der Physik, Volume **17** (1905).

Eliezer, C. J., Gray, A. (1976): *A Note on the Time-Dependent Harmonic Oscillator*, SIAM Journal of Applied Mathematics, Volume **30**, pp. 463 - 468.

Ernst, F. J. (1968): *New Formulation of the Axially Symmetric Gravitational Field Problem I*, Physical Review, Volume **167**, pp. 1175 - 1178; *New Formulation of the Axially Symmetric Gravitational Field Problem II*, Ibidem, Volume **168**, pp. 1415 - 1417.

Ernst, F. J. (1971): *Exterior Algebraic Derivation of Einstein Field Equations Employing a Generalized Basis*, Journal of Mathematical Physics, Volume **12**, pp. 2395 - 2397.

Ertel, H. (1942): (a) *Ein neuer hydrodynamischer Erhaltungssatz*, Naturwissenschaften, Band **30**, pp. 543 - 544; (b) *Ein neuer hydrodynamischer Wirbelsatz*, Meteorologische Zeitschrift, Band **59**, pp. 277 - 281; (c) *Über das Verhältnis des neuen hydrodynamischen Wirbelsatzes zum Zirkulationssatz von V. Bjerknes*, Meteorologische Zeitschrift, Band **59**, pp. 385 - 387; (d) *Über hydrodynamische Wirbelsätze*, Physikalische Zeitschrift, Band **43**, pp. 526 - 529.

Fairlie, D.B. (1999): *Dirac-Born-Infeld equations*, Physics Letters B, Volume **456**, pp. 141 - 146.

Felsager, B. (1983): *Geometry, Particles and Fields*, Odense University Press.

Felsager, B., Leinaas, J. M. (1980): *Geometric Interpretation of Magnetic Fields, and the Motion of Charged Particles*, Nuclear Physics B, Volume **166**, pp. 162 188.

Fennelly, A. J. (1974): *Gravitational Charge, Hadron Hydrodynamics and Gödelized Hadrons*, Nature, Volume **248**, pp. 221 - 223.

Fermi, E. (1950): *High Energy Nuclear Events*, Progress of Theoretical Physics (Japan), Volume **5**, pp. 570 - 583.

Fessenden, R. A. (1900): (a) *A Determination of the Nature of the Electric and Magnetic Quantities and of the Density and Elasticity of the Ether I*, Physical Review, Volume **10**, pp. 1 - 33; (b) *A Determination of the Nature of the Electric and Magnetic Quantities and of the Density and Elasticity of the Ether II*, Physical Review, Volume **10**, pp. 83 - 115.

Feynman, R. P. (1948): *Space-Time Approach to Non-Relativistic Quantum Mechanics*, Reviews of Modern Physics, Volume **20**, pp. 367 - 387.

Feynman, R. P. (1949): *The Theory of Positrons*, Physical Review, Vol. **76**, pp. 749 - 759.

Feynman, R. P. (1965): *The Development of the Space-Time View of Quantum Electrodynamics*, Nobel Lecture, December 11, 1965.

Feynman, R. P. (1969): *Very High-Energy Collisions of Hadrons*, Physical Review Letters, Volume **23**(23), pp. 1415 - 1417.

Feynman, R. P. (1995): *Lectures on Gravitation*, F. B. Morinigo, W. G. Wagner & B. Hatfield Editors, Addison-Wesley Publishing Company, Reading, Massachusetts.

Feynman, R. P., Hibbs, A. R (1965): *Quantum Mechanics and Path Integrals*, McGraw-Hill Publishing Company, New York, NY.

Fisher, R. A. (1922): *On the Mathematical Foundations of Theoretical Statistics*, Philosophical Transactions of the Royal Society of London, Series A, Volume **222**, pp. 309 - 368.

Fisher, R. A. (1925): a) *Applications of "Student's" Distribution*, Metron, Volume **5**, pp. 90 - 104; b) *Theory of Statistical Estimation*, Mathematical Proceedings of the Cambridge Philosophical Society, Volume **22**(4) pp. 700 - 725.

Fixsen, D. J., Cheng, E. S., Gales, J. M., Mather, J. C., Shafer, R. A., Wright, E. L. (1996): *The Cosmic Microwave Background Spectrum from the Full COBE FIRAS Data Set*, The Astrophysical Journal, Volume **473**, pp. 576 - 587.

Flanders, H. (1973): *Differentiation Under the Integral Sign*, The American Mathematical Monthly, Volume **80**(5), pp. 615 - 627.

Fock, V. A. (1959): *The Theory of Space, Time and Gravitation*, Pergamon Press, NY.

Fowles, G. R. (1977): *Self-Inverse Form of the Lorentz Transformation*, American Journal of Physics, Volume **45**(6), pp. 675 - 676.

Frederick, C. (1976): *Stochastic Space-Time and Quantum Theory*, Physical Review D, Volume **13**(11), pp. 179 - 196.

Fresnel, A. (1827): *Mémoire sur la Double Réfraction*, Mmoirs de l'Acadmie des Sciences de l'Institute de France, Tome **7**, pp. 45 - 176; reproduced in Oeuvres Complètes, Imprimerie Impériale, Paris 1866, (can be found at Gallica.bnf.fr)

Friedberg, R., Lee, T. D., Pang, Y., Ren, H. C. (1996): *A Soluble Gauge Model with Gribov- Type Copies*, Annals of Physics, Volume **246**, pp. 381 - 445.

Frieden, B. R. (2004): *Science from Fisher Information*, Cambridge University Press, Cambridge, UK.

Fulton, D. G., Rainich, G. Y. (1932): *Generalizations to Higher Dimensions of the Cauchy Integral Formula*, American Journal of Mathematics, Volume **54**, pp. 235 241.

Gardner, C. S., Greene, J. M., Kruskal, M. D., Miura, R. M. (1967): *Method for Solving the Korteweg-de Vries Equation*, Physical Review Letters, Volume **19**(18), pp. 1095 - 1097.

Georgescu-Roegen, N. (1971): *On the Texture of the Arithmetical Continuum*, Appendix A in *The Entropy Law and the Economic Process*, Harvard University Press, Cambridge MA, USA.

Gheorghiu, Gh. Th. (1964): *Diffferential Geometry*, Editura Didactică şi Pedagogică, Bucharest (in Romanian).

Gibbons, G. W. (1998): *Born-Infeld particles and Dirichlet p-Branes*, Nuclear Physics B, Volume **514**(3), pp. 603 - 639.

Gibbs, J. W. (1883): *On the General Equations of Monochromatic Light in Media of Every Degree of Transparency*, American Journal of Science, Volume **25**, pp. 107 - 118; reproduced in The Scientific Papers of J. Willard Gibbs, Longmans, Green, and Co, New York 1906, Volume II, pp. 211 222.

Gödel, K. (1949): *An Example of a New Type of Cosmological Solutions of Einstein Field Equations of Gravitation*, Reviews of Modern Physics, Volume **21**, pp. 447 - 450; reproduced in General Relativity and Gravitation, Volume **32**, (2000) pp. 1409 1417; *A Remark about the Relationship Between Relativity Theory and Idealistic Philosophy*, n *Albert Einstein, Philosopher-Scientist*, P. A. Schilpp editor; Volume VII, the Library of Living Philosophers, pp. 557 - 562.

Gödel, K. (1952): *Rotating Universes in the General Relativity*, Proceedings of the International Congress of Mathematicians, Cambridge, MA 1950; American Mathematical Society, 1952, Volume 1, pp. 175 181; reproduced in General Relativity and Gravitation, Volume **32**(6), 2000, pp. 1419 - 1427.

Goenner, H. F. M. (2004): *On the History of Unified Field theories I*, Living Reviews in Relativity, Volume **7**, pp. 2 - 153.

Goenner, H. F. M. (2014): *On the History of Unified Field theories II*, Living Reviews in Relativity, Volume **17**, pp. 5 - 241.

Goldman, R. (2005): *Curvature Formulas for Implicit Curves and Surfaces*, Computer Aided Geometric Design, Volume **22**, pp. 632 - 658.

Gradshteyn, I. S., Ryzhik, I. M. (2007): *Table of Integrals, Series and Products*, Seventh Edition, A. Jeffrey & D. Zwillinger Editors, Academic Press-Elsevier.

Grassmann, H. G. (1853): *Zur Theorie der Farbenmischung*, Annalen der Physik und Chemie, Volume **165**(5), pp. 69 - 84; English translation: *On the Theory of Compound Colours*, Philosophical Magazine, Volume 7, pp. 254 264, (1854); reproduced in (MacAdam, 1970).

Greenberger, D. M. (1980): *Comment on "Nonspreading Wave Packets"*, American Journal of Physics, Volume **48**(3), p. 256.

Grigorenko, A. N. (1990): *Geometrical Phase and Thermodynamics*, Physics Letters A, Volume **147**, pp. 427 - 429.

Guggenheimer, H. W. (1977): *Differential Geometry*, Dover Publications, New York.

Halphen, G-H. (1877): *Sur les Lois de Kepler*, Comptes Rendus de l'Académie des Sciences, Paris, Tome **84**, pp. 939 - 941.

Halphen, G. (1879): *Sur l'Équation Différentielle des Coniques*, Bulletin de la Societé Mathématique de France, Tome **7**, pp. 83 - 85.

Hamilton, W. R. (1841): *On a Mode of Deducing the Equation of Fresnel's Wave*, Philosophical Magazine, Volume **19**, pp. 381 - 383; Reproduced in

"The Mathematical Papers of Sir William Rowan Hamilton", A. W. Conway & J. L. Synge Editors, Cambridge University Press 1931, Volume **I**, pp. 341 - 343.

Hannay, J. H. (1985): *Angle Variable Holonomy in Adiabatic Excursion of an Integrable Hamiltonian*, Journal of Physics A: Mathematical and General Volume **18**(2), pp. 221 - 230.

Harvey, R. J. (1966): *Navier-Stokes Analog of Quantum Mechanics*, Physical Review, Volume **152**, p. 1115.

Hasse, R. W. (1978): *Approaches to Nuclear Friction*, Reports on Progress in Physics, Volume **41**, pp. 1027 - 1101.

van Heerden, P. J. (1956): *Concept of Distance in Affine Geometry*, Journal of the Optical Society of America, Volume **46**(11), p. 1000.

Heisenberg, W. (1925): *Über quantentheoretische Umdeutung kinematischer und mechanischer Beziehungen*, Zeitschrift für Physik, Volume **33**, pp. 879 - 893.

Helgason, S. (1984): *Groups and Geometric Analysis*, Academic Press, Inc., NY.

Helmholtz, H. von (1852): *On the Theory of Compound Colours*, Philosophical Magazine, Volume **4**, pp. 519 - 534.

Helmholtz, H. von (1891): (a) *Versuch einer erweiterten Anwendung des Fechner'schen Gesetzes im Farbensystem*, Zeitschrift für Psychologie der Sinnesorgane, Band **2**, pp. 1 - 30; (b) *Versuch das psychophysische Gesetz auf die Farbenunterschiede trichromatischer Augen anzuwenden*, Ibidem, Band **3**, pp. 1 - 20; (c) *Kürzeste Linien im Farbensystem*, Sitzungsberichte der Akademie der Wissenschaften zu Berlin 1891, pp. 1071 - 1083.

Herrero, A., Morales, J. A. (1999): *Radial Conformal Motions in Minkowski Space-Time*, Journal of Mathematical Physics, Volume **40**(6), pp. 3499 - 3508.

Herrero, A., Morales, J. A. (2010): *Painlevé-Gullstrand Synchronizations in Spherical Symmetry*, Classical and Quantum Gravity, Volume **27**, 175007.

Hertz, H. (2003): *The Principles of Mechanics, Presented in a New Form*, Dover Phoenix Editions.

Hoffman, W. C. (1966): *The Lie Algebra of Visual Perception*, Journal of Mathematical Psychology, Volume **3**, pp. 65 - 98.

't Hooft, G (1993): *Dimensional Reduction in Quantum Gravity*, In *Salamfestschrift: a collection of talks*, World Scientific Series in 20th Century Physics, volume **4**, A. Ali, J. Ellis and S. Randjbar-Daemi, Eds., (World Scientific, 1993); **arXiv**: gr-qc/9310026.

Hu, Z.-J., Zhao, G.-S. (1997): *Isometric Immersions of the Hyperbolic Space $H^2(-1)$ Into $H^3(-1)$*, Proceedings of American Mathematical Society, Volume **125**, pp. 2693 - 2697.

Hughes, J. F. (2003): *Differential Geometry of Implicit Surfaces in 3-Space – A Primer*, Report CS-03-05, Department of Computer Science, Brown University, Providence, Rhode Island.

Humbert, P. (1929): (a) *Potentiel Correspondant à une Attraction Proportionnelle à $\rho \cdot exp(\rho^2/2)$*, Mathematica, Cluj-Napoca, Volume **1**, pp. 117121; (b) *Sur une Généralisation de l'Équation de Laplace*, Journal de Mathématiques Pures et Appliques, Tome **8**, pp. 145 - 159.

Israel, W., Wilson, G. A. (1972): *A Class of Stationary Electromagnetic Vacuum Fields*, Journal of Mathematical Physics, Volume **13**, pp. 865 - 867.

Jackiw, R. (1988): *Three Elaborations on Berry's Connection, Curvature and Phase*, International Journal of Modern Physics A, Volume **3**, pp. 285 - 297.

Jackiw, R. (2002): *Physical Instances of Noncommuting Coordinates*, Nuclear Physics B (Proceedings Supplement), Volume **108**, pp. 30 - 36.

Jackiw, R. (2004): *Noncommuting Fields and non-Abelian Fluids*, Nuclear Physics B (Proceedings Supplement), Volume **127**, pp. R327 - R432.

Jackiw, R., Nair, V. P., Pi, S.-Y., Polychronakos, A. P. (2004): *Perfect Fluid Theory and its Extensions*, Journal of Physics A: Mathematical and General, Volume **37**, pp. R327 - R432.

Judd, D. B. (1940): *Hue, Saturation, and Lightness of Surface Colors with Chromatic Illumination*, Journal of Research of the National Bureau of Standards, Volume **24**(3), pp. 293 - 333.

Kanai, E. (1948): *On the Quantization of Dissipative Systems*, Progress of Theoretical Physics, Volume **3**(4), pp. 440 - 442.

Klein, G., Mulholland, H. P. (1978): *Repeated Elastic Reflections of a Particle in an Expanding Sphere*, Proceedings of the Royal Society of London A, Volume **361**, pp. 447 - 461.

Kostin, M. D. (1972): *On the Schrödinger-Langevin Equation*, Journal of Chemical Physics, Volume **57**(9), pp. 3589 - 3591.

Kravtsov, Yu. A. (1988): *Rays and Caustics as Physical Objects*, in *Progress in Optics*, Volume **26**, E. Wolf editor, Elsevier Science Publishers, pp. 228 - 348.

Kravtsov, Yu. A., Orlov, Yu. I. (1982): *Conditions for Applicability of Geometric Diffraction Theory*, Radiophysics and Quantum Electronics **25**, pp. 582 - 590.

Landau, L. D. (1965): *On Multiple Production of Particles during Collisions of Fast Particles*, in Collected Papers, D. Ter Haar editor, Gordon & Breach, Ltd., New York, pp. 569 - 585.

Larmor, J. W. (1900): *On the Statistical Dynamics of Gas Theory as Illustrated by Meteor Swarms and Optical Rays*, Nature, Volume **63**(1624), pp. 168 - 169; British Association Report, September 1900, pp. 632 - 634.

Lax, P. D. (1968): *Integrals of Nonlinear Equations of Evolution and Solitary Waves*, Communications on Pure and Applied Mathematics, Volume **21**(4), pp. 467 - 490.

Lebedeff, P. (1900): *Les Forces de Maxwell-Bartoli dues a la Pression de la Lumière*, Rapports Présentés au Congrès International de Physique, Ch-Éd. Guillaume & L. Poincaré, Eds, Volume II, pp. 133 - 140, Gauthier-Villars, Paris.

Lebedew, P. (1902): *Experimental Investigations on the Pressure of Light*, Astrophysical Journal, Volume **15**, pp. 60 - 61.

Letac, G. (1989): *Le Problème de la Classification des Familles Exponentielles Naturelles de R^d ayant une Fonction-Variance Quadratique*, in *Probability Measures on Group IX*. Lecture Notes in Mathematics, Volume 1379, pp. 194 - 215, Springer, Berlin.

Letac, G. (2016): *Associated Natural Exponential Families and Elliptic Functions*, in *The Fascination of Probability, Statistics and their Applications*, in Honour of Ole E. Barndorff-Nielsen, M. Podolskij & Al, Editors, pp. 53 - 83, Springer International Publishing Switzerland, 2016.

Levi-Civita, T. (1916): *Nozione di Parallelismo in una Varietà Qualunque...*, Rendiconti del Circolo Matematico di Palermo, Volume **42**, pp. 173 - 204.

Lewis, G. N., (1926): *The Conservation of Photons*, Nature, Volume **118**, pp. 874 - 875.

Lorentz, H. A. (1904): *Electromagnetic Phenomena in a System Moving with any Velocity Smaller than that of Light*, Proceedings of the Royal Netherlands Academy of Arts and Sciences at Amsterdam, Volume **6**, pp. 809 - 831.

Lorentz, H. A. (1916): *Theory of Electrons*, Teubner, Leipzig, New York.

Lorenz, L. (1867): *On the Identity of the Vibrations of Light with Electrical Currents*, Philosophical Magazine, Volume **84**, Series 4, pp. 287 - 301 (English translation of the German original from Poggendorff's Annalen, June 1867).

Lowe, P. G. (1980): *A Note on Surface Geometry with Special Reference to Twist*, Math. Proceedings of the Cambridge Philosophical Society, Volume **87**, pp. 481 - 487.

Lummer, O., Pringsheim. E. (1899): *Die Vertheilung der Energie im Spectrum des schwarzen Körpers*, Verhandlungen der Deutschen Physikalischen Gesellschaft im Jahre 1899, pp. 23 - 41.

Lynden-Bell, D. (2006): *Hamilton Eccentricity Vector Generalized to Newton Wonders*, The Observatory, Volume **126**, pp. 176 - 182; **arXiv**:astro-ph/0604428v1,

MacAdam, D. L (1970): *Sources of Color Science*, The MIT Press, Cambridge, MA & London, UK.

Mac Gregor, M. H. (1992): *The Enigmatic Electron*, Kluwer Academic Publishers, Dordrecht, Boston, London.

Mackinnon, L. (1978): *A Nondispersive de Broglie Wave Packet*, Foundations of Physics, Volume **8**(3/4), pp. 157 - 176.

Madelung, E. (1927): *Quantentheorie in hydrodynamischer Form*, Zeitschrift für Physik, Volume **40**, pp. 322 - 326; *Quantum Theory in Hydrodynamical Form*, translation by D. H. Delphenich; www.neo-classical-physics.info.

Madore, J. (1991): *The Fuzzy Sphere*, Preprint LPHTE, Orsay 91/09; Classical and Quantum Gravity, Volume **9**, pp. 69 - 80 (1992).

Madore, J. (1992): *Fuzzy Physics*, Preprint LPHTE, Orsay 92/01; Annals of Physics, Volume **219**, pp. 187 - 198.

Mandel, L. (1974): *Interpretation of Instantaneous Frequencies*, American Journal of Physics, Volume **42**, pp. 840 - 848.

Manton, N. S., Sutcliffe, P. M. (2004): *Topological Solitons*, Cambrige University Press, Cambridge, UK.

Martinov, D., Ouroushev, D., Grigorov, A. (1991): *About One Superposition of Solutions of the Laplace Equation*, Journal of Physics A: Mathematical and General, Volume **24**, pp. L975 - L979.

Martinov, D., Ouroushev, D., Grigorov, A. (1992): *New Class Solutions of the 3D Laplace Equation*, Journal of Mathematical Physics, Volume *33*, pp. 822 - 825.

Mazilu, N. (2006): *The Stoka Theorem, a Side Story of Physics in Gravitation Field*, Supplemento ai Rendiconti del Circolo Matematico di Palermo **77**, pp. 415 - 440.

Mazilu, N. (2010): *Black-Body Radiation Once More*, Bulletin of the Polytechnic Institute of Jassy, Volume **54**, pp. 69 - 97.

Mazilu, N., Agop, M. (2012): *Skyrmions − a Great Finishing Touch to Classical Newtonian Philosophy*, Nova Publishers, New York.

Mazilu, N., Agop, M. (2015): *The Concept of Physical Surface in Nuclear Matter*, Modern Physics Letters A, Volume **30**(5), 1550026.

Mazilu, N., Porumbreanu, M. (2018): *Devenirea Mecanicii Ondulatorii*, Editura Limes, Cluj- Napoca (soon to appear in English as *The Coming to Being of Wave Mechanics*).

Mercheş, I., Agop, M. (2015): *Differentiability and Fractality in Dynamics of Physical Systems*, World Scientific, Singapore.

Milnor, J. (1983): *On the Geometry of the Kepler Problem*, The American Mathematical Monthly, Volume **90**(6), pp. 353 - 365.

Misner, C. W. (1968): *The Isotropy of the Universe*, The Astrophysical Journal, Volume 151(2), pp. 431 - 457.

Misner, C. W. (1978): *Harmonic Maps as Models for Physical Theories*, Physical Review D, Volume **18**, pp. 4510 - 4524.

Misner, C. W., Thorne, K. S., Wheeler, J. A. (1973): *Gravitation*, W. H. Freeman and Company, San Francisco, CA.

von Mises, R. (1913): *Mechanik der festen Körper im plastisch-deformablen Zustand*, Nachrichten der Königlichen Gesellschaft der Wissenschaften zu Göttingen. Mathematisch-physikalische Klasse, 1913, pp. 582 - 592; *Mechanics of solid bodies in the plastically-deformable state*, English version, due to D. H. Delphenich, of the German original; one can find it at www.neo-classical-physics.info.

Mittag, L., Stephen, M.J. (1992): *Conformal Transformations and the Application of Complex Variables in Mechanics and Quantum Mechanics*, American Journal of Physics, Volume **60**, pp. 207 - 211.

Nedeff, V., Lazăr, G., Agop, M., Eva, L., Ochiuz, L., Dimitriu, D., Vrăjitoriu, L., Popa, C. (2015): *Solid Components Separation from Heterogeneous Mixtures through Turbulence Control*, Powder Technology, Volume **284**, pp. 170 - 186.

Nelson, E. (1966): *Derivation of the Schrödinger Equation from Newtonian Mechanics*, Physical Review, Volume **150**, pp. 1079 - 1085.

Newton, Sir Isaac (1952): *Opticks, or a Treatise of the Reflections, Refractions, Inflections & Colours of Light*, Dover Publications, Inc., New York.

Newton, I. (1974): *The Mathematical Principles of Natural Philosophy*, Translated into English by Andrew Motte in 1729. The translations revised, and supplied with an historical and explanatory appendix, by Florian Cajori, University of California Press, Berkeley and Los Angeles, California.

Niall, K. K. (2017): *Erwin Schrödinger's Color Theory*, Translated with Modern Comentary, Springer International Publishing AG.

Niederer, U. (1972): *The Maximal Kinematical Invariance Group of the Free Schrödinger Equation*, Helvetica Physica Acta, Volume **45**, pp. 802 - 810.

Nottale, L. (1992): *The Theory of Scale Relativity*, International Journal of Modern Physics A, Volume **7**(20), pp. 4899 - 4936; the www version complemented by notes and errata in 2003.

Nottale, L. (2011): *Scale Relativity and Fractal Space-Time*, Imperial College Press, London, UK.

Novozhilov, V. V. (1952): *On the Physical Meaning of the Invariants Used in the Theory of Plasticity (in Russian)*, Prikladnaya Matematika i Mekhanika, Volume **16**, pp. 617 - 619.

Olver, P. J. (1998): *Applications of Lie Groups to Differential Equations*, Springer-Verlag New York, Inc.

Ouroushev, D. (1985): *Three-Dimensional Analytical Periodic Solutions of the Laplace Equation*, Journal of Physics A: Mathematical and General **18**, pp. L845 - L848.

Penzias, A. A., Wilson, R. W. (1965): *A Measurement of Excess Antenna Temperature at 4080 Mc/s*, The Astrophysical Journal, Volume **142**, pp. 419 - 421.

Perjés, Z. (1970): *Spinor Treatment of Stationary Space-Times*, Journal of Mathematical Physics, Volume **11**, pp. 3383 - 3391.

Pissondes, J.C. (1999): *Covariance in General Relativity and Scale-Covariance in Scale- Relalivity Theory, Quadratic Invariants and Leibniz Rule*, Chaos, Solitons & Fractals, Volume **10**(2-3), pp. 513 - 541.

Planck, M (1900): *Planck's Original Papers in Quantum Physics*, translated by D. Ter-Haar and S. G. Brush, and Annotated by H. Kangro, Wiley & Sons, NY 1972.

Poincaré, H (1895): *Capillarité*, Georges Carré, Paris.

Poincaré, H. (1896): *Remarques sur une Expérience de M. Birkeland*, Comptes Rendus de l'Académie des Sciences de Paris, Tome **123**, pp. 530 - 533.

Poincaré, H. (1897): *Les Idées de Hertz sur la Mécanique*, Revue Générale des Sciences Pures et Appliquées, Tome **8**(17), pp. 734 - 743 (there is an English translation of this work at vixra.org).

Pokorny, J., Smith, V. C., Xu, J. (2012): Quantal and Non-Quantal Color Matches: Failure of Grassmann's Laws at Short Wavelengths, Journal of the Optical Society of America A, Volume **29**(2), pp. A324 - A336.

Priest, I. G. (1919): (a) *A One-Term Pure Exponential Formula for the Spectral Distribution of Radiant Energy from a Complete Radiator*, Journal of the Optical Society of America, Volume **2**, pp. 1822; (b) *A New Formula for the Spectral Distribution of Energy from a Complete Radiator*, Physical Review, Volume 13, pp. 314 - 317.

Rainich, G. Y. (1925): *Sur une Répresentation des Surfaces*, Comptes Rendus de l'Académie de Sciences, Paris, Tome **180**, pp. 801 - 803.

Reddy, J. N. (2008): *An Introduction to Continuum Mechanics, with Applications*, Cambridge University Press, Cambridge UK.

Regge, T. (1959): *Introduction to Complex Orbital Momenta*, Il Nuovo Cimento, Volume **14**(5), pp. 951 - 976.

Renn, J. (General Editor) (2007): *The Genesis of General Relativity*, Four-volume Compendium, forming Volume **250** of Boston Studies in the Philosophy of Science, Springer, New York.

Resnikoff, H. L. (1974): *Differential Geometry of Color Perception*, Journal of Mathematical Biology, Volume **1**, pp. 97 - 131.

Reynolds, O. (1903): *The Sub-Mechanics of the Universe*, Cambridge University Press, Cambridge UK.

Riemann, B. (1867): *Ueber die Hypothesen, welche der Geometrie zu Grunde liegen*, Abhandlungen der Königlichen Gesellschaft der Wissenschaften zu Göttingen, Volume **13**, pp. 2082 - 2089.

Rosen, G. (1972): *Galilean Invariance and the General Covariance of Nonrelativistic Laws*, American Journal of Physics, Volume **40**(5), pp. 683 -

687.

Rosen, N. (1945): *On Waves and Particles*, Journal of Elisha Mitchell Scientific Society, Volume **61**, pp. 67 - 73.

Rosen, N. (1947): *Statistical Geometry and Fundamental Particles*, Physical Review, Volume **72**, pp. 298 - 303.

Russell, J. M. (2013): *Applications of Kelvin's Inversion Theorem to the Solution of Laplace's Equation over a Domain that Includes the Unbounded Exterior of a Sphere*, Proceedings of the 2013 COMSOL Conference in Boston MA.

Salmon, R. (1988): *Hamiltonian Fluid Mechanics*, Annual Reviews in Fluid Mechanics, Volume **20**, pp. 225 - 256.

Salmon, R. (1998): *Lectures on Geophysical Fluid Dynamics*, Oxford University Press, Oxford, UK.

Sasaki, R. (1979): *Soliton Equations and Pseudospherical Surfaces*, Nuclear Physics B, Volume **154**, pp. 343 - 357.

von Schelling, H. (1956): *Concept of Distance in Affine Geometry and Its Applications in Theories of Vision*, Journal of the Optical Society of America, Volume **46** (5), pp. 309 - 315; *Reply to P. J. van Heerden*, Ibidem 46(11), pp. 1000 - 1001.

Schleich, W. P., Greenberger, D. M., Kobe, D, H., Scully, M. O. (2013): *Schrödinger Equation Revisited*, Proceedings of the National Academy of Sciences USA, Volume *110*(14), pp. 5374 - 5379.

Schrödinger, E. (1920): *Grundlinien einer Theorie der Farbenmetrik im Tagessehen*, Annalen der Physik, Band **63**(21), pp. 397 - 426; 427 - 456; **63**(22), pp. 481 - 520; English translations by Keith K Niall, in the collection (Niall, 2017).

Schrödinger, E. (1933): *Mémoires sur la Mécanique Ondulatoire*, Librairie Félix Alcan.

Schubert, W., Ruprecht, E., Hertenstein, R., Nieto Ferreira, R., Taft, R., Rozoff, C., Ciesielski, P., Kuo, H.-C. (2004): *English Translations of Twenty-One of Ertel's Papers on Geophysical Fluid Dynamics*, Meteorologische Zeitschrift, Volume **13**(5), pp. 527 - 576; these translations — together with a few other fundamental works of geophysics, either written directly or translated into English — are at the disposal of those interested in the personal page of Professor Geoffrey K. Vallis, from the Department of Mathematics of Exeter University, UK.

Scott, A. C., Chu, F. Y. F., McLaughlin, D. W. (1973): *The Soliton: A New Concept in Applied Science*, Proceedings of the IEEE, Volume **61**(10), pp. 1443 - 1483.

Shapere, A., Wilczek, F. (1989): *Geometric Phases in Physics*, World Scientific Publishing Co., Singapore.

Shchepetilov, A. V. (2003): *The Geometric Sense of the Sasaki Connection*, Journal of Physics A: Mathematical and General, Volume **36**(12), pp. 3893 - 3898.

Simon, B. (1983): *Holonomy, the Quantum Adiabatic Theorem, and Berry's Phase*, Physical Review Letters, Volume **51**(23), pp. 2167 - 2170.

Skinner, D. (2016): *Mathematical Methods*, Course Notes at the University of Cambridge, UK.

Shpilker, G. L. (1984): *The Correct Reconstruction of Fields Satisfying D'Alembert Equation* (in Russian), Doklady, Volume **264**, pp. 1355 - 1358; translated in English by Earth Sciences Section, American Geological Institute.

Spivak, M. (1995): *Calculus on Manifolds*, Addison-Wesley, Reading, Massachusetts.

Spivak, M. (1999): *A Comprehensive Introduction to Differential Geometry*, Volume III, Publish or Perish, Inc., Houston, Texas.

Stoka, M. I. (1968): *Géométrie Intégrale*, Mémorial des Sciences Mathématiques, Fascicule 165, pp. 1 - 65, Gauthier Villars, Paris.

Stoker, J. J. (1989): *Differential Geometry*, Wiley Interscience, New York.

Stratton, J. A. (1941): *Electromagnetic Theory*, McGraw-Hill Book Company, Inc., NY.

Struik, D. J. (1988): *Lectures on Classical Differential Geometry*, Dover, NY.

Susskind, L. (1994): *The World as a Hologram*, Journal of Mathematical Physics, Volume **36**(11), pp. 6377 - 6396; arxiv: hep-th/9409089.

Švec, A. (1988): *Infinitesimal Rigidity of Surfaces in A^3*, Czechoslovak Mathematical Journal, Volume **38**(3), pp. 479 - 485.

Synge, J. L. (1972): *A Special Class of Solutions of the Schrödinger Equation for a Free Particle*, Foundations of Physics, Volume **2**(1), pp. 35 - 40.

Takabayasi, T (1952): *On the Formulation of Quantum Mechanics associated with Classical Pictures*, Progress of Theoretical Physics, Volume **8**(2), pp. 143 182.

Truesdell, C. A. (1951): (a) *On Ertel's Vorticity Theorem*, ZAMP, Volume **2**(2), pp. 109 - 114; (b) *Proof that Ertel's Vorticity Theorem Holds in Average for any Medium Suffering no Tangential Acceleration on the Boundary*, Geofisica Pura e Applicata, Volume **19**(3-4), pp. 167 - 169.

Vallée, O., Soares, M. (2004): *Airy Functions and Applications to Physics*, World Scientific, Singapore.

Vrânceanu, G. (1962): *Lessons of Differential Geometry*, Volume I, Editura Academiei, Bucureşti (in Romanian).

Whittaker, E. T. (1903): *On the Partial Differential Equations of Mathematical Physics*, Mathematische Annalen, Volume 57, pp. 333 - 355.

Whittaker, E. T. (1917): *Analytical Dynamics*, Cambridge University Press.

Widder, D. V. (1979): *The Airy Transform*, The American Mathematical Monthly, Volume **86**, pp. 271 - 277.

Wien, W. (1896): *Über die Energievertheilung im Emissionsspectrum eines schwarzen Körpers*, Annalen der Physik, Volume **58**, pp. 662 - 669.

Wien, W. (1900): *Les Lois Théoriques du Rayonnement*, Rapports Présentés au Congrès International de Physique, editori Ch-Éd. Guillaume & L. Poincaré, Volume **II**, pp. 23 - 40, Gauthier-Villars, Paris.

Wien, W., Lummer, O. (1895): *Methode zur Prüfung des Strahlüngsgesetzes absolut schwarzer Körper*, Annalen der Physik (Wiedemann), Band **56**, pp. 451 - 456.

Wigner, E. P. (1954): *Conservation Laws in Classical and Quantum Physics*, Progress of Theoretical Physics (Japan), Volume **11**(4/5), pp. 437 - 440.

Wilhelm, H. E. (1971): *Hydrodynamic Model of Quantum Mechanics*, Physical Review, Series D, Volume **1**, pp. 2278 - 2285.

Winkelmolen, A. M. (1982): *Critical Remarks on Grain Parameters, with Special Emphasis on Shape*, Sedimentology, Volume **29**(2), pp. 255 - 265.

Wyszecki, G., Stiles, W. S. (1982): *Color Science: Concepts and Methods, Quantitative Data and Formulae*, John Wiley & Sons, New York.

Yamamoto, T. (1952): *The Analytic Representation of Spin*, Progress of Theoretical Physics (Japan), Volume **8**, p. 258.

Yang, C. N., Mills, R. L. (1954): *Conservation of Isotopic Spin and Isotopic Gauge Invariance*, Physical Review, Volume **96**, pp. 191-195.

Zelikin, M. I. (2000): *Control Theory and Optimization*, Encyclopaedia of Mathematical Sciences, Volume **86**, Springer.

Subject Index

A

affine reference frame, 73, 74
Agaokas theorem, 112
Airy moment, 154, 157, 224, 226
Appell coordinates, 124
Appells procedure, 122
Appells transformation, 123
application function
      Louis de Broglie, 152
asymptotic freedom, 15, 32, 101
average density, 6, 8

B

Berry moment, 96, 99, 157, 223
Berry-Klein scaling, 107
Berry-Klein theory, 162, 166
blackbody radiation, 85, 98,
      105, 163
Born interpretation, 18
Born-Infeld electrodynamics, 209

C

cardinality, 78, 227
Cartesian reference frames, 165
compatibility equations
      between metric and
          connection, 208, 210
confinement
      mathematical idea, 71, 102

continuity equation, 13, 23, 30,
      166, 173, 176, 181
control surface, 173, 174
control volume, 170, 172, 179, 182
cosmic background radiation, 98,
      99, 173, 174
control volume, 170, 172, 179, 182
cosmic background radiation, 98,
      99

D

de Broglie moment, 150, 222, 228
de Broglie tube, 50, 58. 102
definition of interpretation, 45, 110
density tensor, 167
distant parallelism, 196, 198, 208,
      227
dynamics, 61, 110, 140, 160, 209,
      244

E

Ehrenfest's theorem, 26, 103, 185
Einstein elevator, 100, 156, 225
Ernst potential, 68, 69
Ertel's theorem, 179, 181, 182